MW00837547

The Mixing of Rubber

The Mixing of Rubber

Edited by

Richard F. Grossman

HALSTAB
Hammond
Indiana
USA

CHAPMAN & HALL
London · Weinheim · New York · Tokyo · Melbourne · Madras

Published by Chapman & Hall, 2–6 Boundary Row, London SE1 8HN, UK

Chapman & Hall, 2–6 Boundary Row, London SE1 8HN, UK

Chapman & Hall GmbH, Pappelallee 3, 69469 Weinheim, Germany

Chapman & Hall USA, 115 Fifth Avenue, New York, NY 10003, USA

Chapman & Hall Japan, ITP-Japan, Kyowa Building, 3F, 2-2-1 Hirakawacho, Chiyoda-ku, Tokyo 102, Japan

Chapman & Hall Australia, 102 Dodds Street, South Melbourne, Victoria 3205, Australia

Chapman & Hall India, R. Seshadri, 32 Second Main Road, CIT East, Madras 600 035, India

First edition 1997

© 1997 Chapman & Hall

Typeset in 10/12 Palatino by Florencetype Ltd, Stoodleigh, Devon
Printed in Great Britain by Hartnolls Ltd, Bodmin, Cornwall

ISBN 0 412 80490 5

Apart from any fair dealing for the purposes of research or private study, or criticism or review, as permitted under the UK Copyright Designs and Patents Act, 1988, this publication may not be reproduced, stored, or transmitted, in any form or by any means, without the prior permission in writing of the publishers or in the case of reprographic reproduction only in accordance with the terms of the licences issued by the Copyright Licensing Agency in the UK, or in accordance with the terms of licences issued by the appropriate Reproduction Rights Organization outside the UK. Enquiries concerning reproduction outside the terms stated here should be sent to the publishers at the London address printed on this page.

The publisher makes no representation, express or implied, with regard to the accuracy of the information contained in this book and cannot accept any legal responsibility or liability for any errors or omissions that may be made.

A catalogue record for this book is available from the British Library

∞ Printed on permanent acid-free text paper, manufactured in accordance with ANSI/NISO Z39.48–1992 and ANSI/NISO Z39.48–1984 (Permanence of Paper).

Contents

Contributors

Allen C. Bluestein
Berlington Associates, Inc.
232 Burlington Ave.
Spotswood, NJ 08884

E. L. Canedo
Farrel Corporation
25 Main St.
Ansonia, CT 06401

Charanjit S. Chodha
Uniroyal Chemical Co.
Spencer St.
Naugatuck, CT 06770

R. J. Del Vecchio
Technical Consulting Services
3 John Matthews Road
Southborough, MA 01772

G. S. Donoian
Farrel Corporation
25 Main St.
Ansonia, CT 06401

Richard F. Grossman
Halstab
1013 Oriente Ave.
Wilmington, DE 19807

Martin J. Hannon
Uniroyal Chemical Co.
Spencer Street
Naugatuck, CT 06770

Emmanuel G. Kontos
Uniroyal Chemical Co.
Spencer St.
Naugatuck, CT 06770

Richard Mastromatteo
Cri-Tech, Inc.
85 Winter St.
Hanover, MA 02339

Michael A. Melotto
Farrel Corporation
25 Main St.
Ansonia, CT 06401

Robert F. Ohm
R. T. Vanderbilt Co., Inc.
30 Winfield St.
Norwalk, CT 06855

Steven R. Salma
Farrel Corporation
25 Main St.
Ansonia, CT 06401

Gene J. Sorcinelli
Farrel Corporation
25 Main St.
Ansonia, CT 06401

L. N. Valsamis
Farrel Corporation
25 Main St.
Ansonia, CT 06401

Preface

What if this mixture do not work at all?
Shakespeare, *Romeo and Juliet*

So says Juliet in Act IV, scene iii; if the mixture is a rubber compound, whether it works or not will depend on several premises:

- That a composition has been developed whose properties, as measured experimentally, give rise to the expectation that it will work.
- That raw materials have been obtained whose properties correspond in adequate measure to those of the samples used in formulation development.
- That the composition may be consistently mixed to suitable homogeneity by those skilled in the art, with efficiency commensurate with the added value commanded in the marketplace.
- That the mixed compound may similarly be fabricated into marketable articles.

The mixing of rubber is concerned mainly with the third of the above considerations. Nevertheless, mixing cannot be isolated from formulation, raw material control and the influence of further processing. These topics are therefore considered in terms of their effects on mixing of rubber compounds.

Only in the tire sector of the rubber industry is mixing itself considered a technical specialty. In most other areas, the manufacturer of rubber articles will generally have strong expertise in further processing: extrusion, calendering, molding, etc. Leading manufacturers will typically invest in the most modern processing equipment. And they will generally invest in acquiring the materials science underlying their specialities. Often, however, they will persist with obsolete mixing equipment and mixing procedures that are historic rather than ingenious.

Innovation in mixing is associated with the tire industry, some equipment manufacturers, certain custom compounders, a small number of manufacturers in specialized fields, and some of their suppliers. Few in any of these categories are willing to discuss their programs with the industry at large. I would like to thank the authors who have, in every case, been able to convince the management of their companies of the long-range benefits of education.

Richard F. Grossman
Wilmington, Delaware
November 1996

Trade names

Altax, Amax, Cumate, Leegen, Tuads, Vanax, Vanfre, Vanox, and Vanplast are registered in the United States to the R. T. Vanderbilt Company, Inc. Banbury, FCM, MVX, and ST (synchronous technology) in reference to internal mixer rotors, are registered to Farrel Corporation. Plioflex and Wingstay are trade names of Goodyear Tire & Rubber Co., Carbowax of Union Carbide Corp., HiSil of Pittsburgh Plate Glass Co., and Renacit of Bayer. Hypalon, Vamac, and Viton are registered by E. I. duPont de Nemours, Thiokol by Morton International, Factice by Harwick Corp., AgeRite and Stalite by the B. F. Goodrich Company, Royalene by Uniroyal Chemical Co., Pepton by Shaw Chemical, and Dyphos by Halstab Division, The Hammond Group.

Acronyms

ACN	acrylonitrile
BR	butadiene rubber
CP	chemical peptizer
CPE	chlorinated polyethylene
CR	chloroprene rubber
CSM	chlorosulfonated polyethylene
CV	continuous vulcanization
DBP	dibutyl phthalate
DCPD	dicyclopentadiene
DOP	di-2-ethylhexyl phthalate
ECO	epichlorohydrin copolymer
ENB	ethylidene norbornene
EPDM	ethylene–propylene–diene terpolymer
EPM	ethylene–propylene copolymer
ETU	ethylene thiourea
EVA	ethylene-vinyl acetate copolymer
FCM	Farrel continuous mixer
FEF	fast extruding furnace black, ASTM N500 series
FKM	fluoroelastomer
GPF	general-purpose furnace black, ASTM N600 series
HAF	high abrasion furnace black, ASTM N300 series
HD	hexadiene
ILPA	internally lubricating processing aid
IR	isoprene rubber
ISAF	intermediate super abrasion furnace black, ASTM N200 series
LDPE	low density polyethylene
MBT	mercaptobenzothiazole
MBTS	benzothiazyl disulfide
MPT	Monsanto processability tester
MW	molecular weight

MWD	molecular weight distribution
NBR	nitrile–butadiene rubber
NPD	number of passage distribution
NR	natural rubber
ODR	oscillating disk rheometer
PE	polyethylene
phr	parts per hundred rubber
PPDC	piperidinium pentamethylene dithiocarbamate
PTFE	polytetrafluoroethylene
PVC	polyvinyl chloride
QC	quality control
SBR	styrene–butadiene rubber
SCR	silicon-controlled rectifier
SPC	statistical process control
SRF	semireinforcing furnace black, ASTM N700 series
TBBS	*t*-butylbenzothiazole sulfenamide
TETD	tetraethylthiuram disulfide
TMTD	tetramethylthiuram disulfide
TPE	thermoplastic elastomer
TQM	total quality management
XLPE	cross-linkable polyethylene

1

Mixing machinery for rubber

Michael A. Melotto

1.1 INTRODUCTION

This chapter will deal broadly with the art and science of mixing rubber. Those unfamiliar with the nuts-and-bolts aspect of the industry may all too readily attribute artistic (or even magical) practice to the exotic ingredients and process techniques involved in manufacturing rubber articles. Others who have experienced the frustration when a composition does not quite meet a required physical property, lacks an anticipated attribute, or processes unsatisfactorily for no apparent reason, only to uncover remedies that are as difficult to explain as the symptoms, they will understand the phrase 'art and science'. These occurrences do not detract from the scientific achievements that have driven the art of mixing to its current level of sophistication.

In order to understand the reasons for the techniques and types of machinery employed in mixing, one must have some familiarity with raw materials, their physical forms, functions in the compound, and behavior during processing. Several basic categories of ingredients are usually distinguished:

- **Rubber or polymer**: bales, chips, pellets, or powder
- **Fillers**: powder, pellets
  - Reinforcing: carbon black, silica
  - Extending: clay, calcium carbonate, talc
- **Plasticizers and lubricants**: fluids, oils, waxes
  - Process oil, ester plasticizers
  - Processing aids: waxes, proprietary blends; stearic acid

The Mixing of Rubber
Edited by Richard F. Grossman
Published in 1997 by Chapman & Hall, London. ISBN 0 412 80490 5.

- **Miscellaneous additives**: powder, pellets, fluids
  - Antioxidants, antiozonants
  - Colorants
  - Tackifiers, release agents
- Vulcanizing agents and accelerators
  - Sulfur
  - Peroxides
  - Special ingredients

A knowledge of physical characteristics and forms (elaborated in later chapters) leads to a certain amount of guidance regarding the most suitable type of mixer. For example, the knowledge that liquid additives need to be mixed into a liquid polymer would direct the technologist to certain types of equipment. But selection of the optimum mixer is not always possible, primarily because of economic factors. The organization must often make do with what happens to be available; the ingenuity of the compounder lies in his or her ability to achieve specific results with the equipment on hand. On the other hand, given some degree of choice, a more appropriate machine is usually available to improve mixing characteristics and therefore final properties.

Early in the history of rubber compounding, almost simultaneously with Goodyear's discovery of sulfur vulcanization, it was found that kneading, or softening, the elastomer was useful in increasing its receptivity to incorporation of powders. This is the basis of mixing – masticating the elastomer to make it receptive to other ingredients, yet retaining sufficient stiffness to ensure adequate dispersion. (The more difficult an ingredient is to disperse, the higher the viscosity required during mixing.) This balance of mastication without undue shear softening can be achieved through several means:

1. Mixing with close temperature control
2. Use of a specific sequence for adding materials
3. In some cases, remixing after cooling (two-pass mixing)

These three considerations apply to mixing with all the most common types of equipment: two-roll mills, internal batch mixers, continuous mixers, extruders or combinations thereof.

1.2 TWO-ROLL MILLS

Every mixer must provide two basic functions, both equally important: acceptable dispersion (intensive or dispersive mixing) and high uniformity (extensive or distributive mixing). In many cases the piece of equipment used most often by the rubber technologist is the two-roll lab mill, (Fig. 1.1 and Table 1.1), a device for preparing small quantities

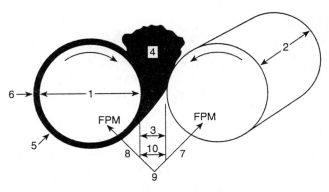

Figure 1.1 Two-roll mill: numbered quantities are defined in Table 1.1. (Courtesy the Farrel Corporation)

Table 1.1 Roll nomenclature: numbers refer to Fig. 1.1

1	Diameter (*D*)	Usually same for both rolls
2	Face length (*L*)	Roll length (mill sizes expressed as *D*×*L*)
3	Roll gap	Distance between rolls
4	Bank size	Material sitting above gap
5	Banded roll	Roll which material follows
6	Front roll	Roll on operator's side
7	Slow roll	Roll rotating at slowest speed
8	Fast roll	Roll rotating at fastest speed
9	Friction ratio	Roll speed ratio
10	Separating force	Resultant force exerted by material in roll gap

of mixed compound. This mixing device is usually set for a ratio of roll surface frictional speed of about 1.25:1.

The rolls are bored to permit cooling or heating; the gap between the rolls is adjustable within a range related to roll diameter. The mixing procedure is relatively standard. The operator places portions of elastomer on the mill, kneading the sample by multiple passes through the gap, until sufficient reduction in stiffness permits it to wrap and adhere to one roll. The gap is adjusted so that a reservoir of elastomer is always rotating above the nip. This reservoir is called the **rolling bank**. Rolling is rarely observed when the polymer has only limited elastomeric character at the milling temperature. In such cases the reservoir may flop about or break into discrete sections. This behavior may often be corrected by a different choice of mill temperature on one roll or both, sometimes merely by using a different nip setting.

Powders are now added into the gap, with frequent pan sweeping to recover increments which drop through. Process oil or plasticizer is also added, usually after part of the filler content has been incorporated. Only then does the mill operator begin to cut the rubber from one side of the mill, passing it to the other side, to crossblend the batch. At the conclusion of mixing (usually determined by the judgment either of the operator or a supervisor) the batch is cut from the mill for cooling and storage. A major advantage of mill mixing is the high shear developed at the mill nip; this breaks up agglomerates and drives incorporation of ingredients. Furthermore, the massive surface exposure imparts good cooling, thereby maintaining the stiffness of the compound. And due to the roll friction ratio, the rolling bank imparts further high shear.

The disadvantages of mill mixing usually far outnumber the advantages and may be listed as follows:

1. Length of the mixing cycles
2. Dependence on operator skills
3. Dust and dirt levels that are typical
4. Difficulty in standardizing subjective procedures
5. Difficulty in controlling batch-to-batch uniformity

Mills are used more in forming and breakdown applications than in actual mixing, except for addition of curatives to premixed masterbatch. This topic will be discussed in greater detail in Chapter 3.

The most common ratio of roll speeds in the past was 1.25:1, with the slower of the two rolls usually on what is most frequently the operator's side (often called the 'front' roll). Recent investigations have shown that the temperature rise of rubber on a two-roll mill is directly related to the sum of the speeds of the two rolls. Therefore, whether this sum is reduced by slowing the faster of the two rolls (a change of friction ratio), or by reducing the speed of both rolls (maintaining the same friction ratio), the result is a reduction in rubber temperature buildup. Mills built in recent years have had lower total speed, ratios closer to 1.1:1, and the fast roll at the front. In addition to improved processing of a broader range of elastomeric compounds, many such mills have also featured drilled rolls, permitting better temperature control and leading to easier compound release.

1.3 INTERNAL BATCH MIXERS

The Banbury internal mixer was originally manufactured to replace the two-roll mill. The original nomenclature indicated the approximate

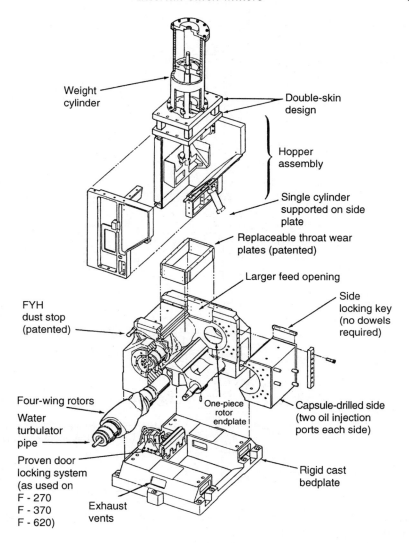

Figure 1.2 Basic Banbury mixer. (Courtesy the Farrel Corporation)

number of 60in mills that a specific Banbury size could equal in output. The basic design of the machine (Fig. 1.2) includes two rotors that operate at a slight speed differential. The rotors are noninterlocking. Mixing or shearing action occurs between the rotors and the sides of the mixer, and between the rotors themselves. The mixer is top loaded through an opening large enough to accommodate bales of elastomers (as well

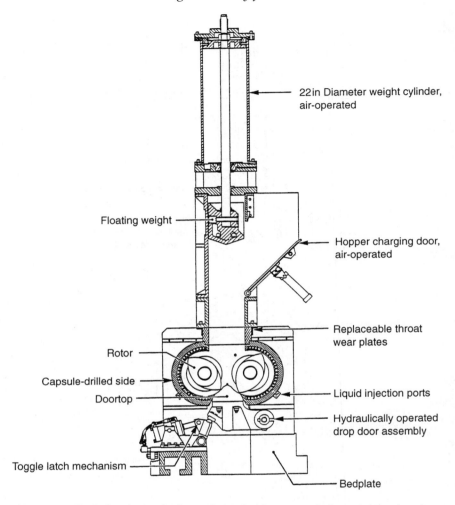

22 in Diameter weight cylinder, air-operated

Floating weight

Hopper charging door, air-operated

Replaceable throat wear plates

Rotor

Capsule-drilled side

Doortop

Liquid injection ports

Hydraulically operated drop door assembly

Toggle latch mechanism

Bedplate

Figure 1.3 Each batch is discharged at the bottom of the mixing chamber. (Courtesy the Farrel Corporation)

as the other ingredients). Pressure is exerted on the batch using a ram which closes the feed opening. Discharge of the batch occurs at the bottom of the mixing chamber (Fig. 1.3). The rotor design is such that material in the chamber is constantly being displaced, corresponding to the crossblending action of the mill operator cutting the batch on a two-roll mill. The compound is subjected to the shearing action of the rotors against the sides, and the action of the rolling bank between the rotors.

The mixing chamber sides and the rotors are cooled to maintain a high rate of shear during the mixing process. Improved cooling has been one of the most significant changes in modification of Banbury mixers over the years. The currently used cooling passages, called **drilled sides**, impart an ease of heat transfer never before obtainable. In fact, the heat transfer efficiency of drilled-side mixers is so great that too much cooling can occur, leading to inefficiency when mixing certain types of compounds. This is corrected, and mixing generally much improved, by the use of a closed-circuit coolant tempering system. This system is used to dial in the proper temperature, optimizing the degree of heat transfer and complementing the viscosity of a specific compound. (Section 9.2 discusses the significance of this feature when using tempered water, an important application.)

As with the two-roll mill, the sequence of ingredient additions is critical. In the case of the high viscosity elastomers, the polymer is usually added and masticated before the fillers are added; then the plasticizers and softeners are finally introduced. If the particular compound contains no curatives, it is then discharged at an appropriate predetermined temperature. If the compound is a single-pass thermosetting composition, the curatives and accelerators are also added and the batch discharged at a lower temperature to prevent scorch (premature curing). The criterion for completion of the batch is often mixing to a specific temperature. Other criteria include elapsed time, and total power input into the batch. A combination of these observations is commonly used to judge the endpoint and related to further processing of the batch during fabrication or to observed properties after vulcanization.

The selection of rotor speeds, mixer size and the extent of materials-handling automation are all related to the demands of the specific sector of the rubber industry. At one extreme there is the massive quantity of compound mixed by tire manufacturers, where most compounds are two-stage mixed (masterbatch and final) using high speed, automated materials-handling equipment, and large mixers. On the other hand, many custom compounders and mechanical goods manufacturers (having large numbers of recipes) prefer lower speeds and smaller machines, and they have too great a variety of materials for totally automated handling. In many cases their compounds are mixed in a single cycle to reduce handling and inventories.

The major advantages of the internal batch mixer are as follows:

1. Highly reproducible cycles
2. Minimum dependence upon operator skills
3. Large capacity and high output
4. Relatively short mixing cycles
5. Potentially clean factory operations

But there are some disadvantages, compared to mill mixing:

1. More rapid temperature rise in mixing
2. More time needed for cleaning equipment
3. Much higher initial investment

1.4 CONTINUOUS MIXERS

Compared to thermoplastics, continuous mixing of rubber compounds is a relatively new concept. The machinery available ranges from simple single-screw extruders to twin-rotor, multistage machines. All of these mixing extruders place a common demand upon materials; the ingredients need to be available in a free-flowing form, capable of being continuously weighed and metered to the mixer. The continuous

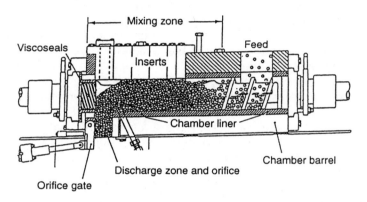

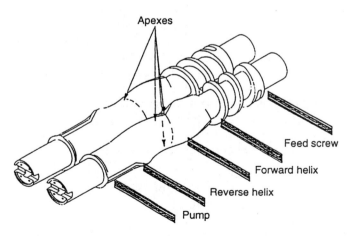

Figure 1.4 Farrel continuous mixer. (Courtesy the Farrel Corporation)

mixer (Fig. 1.4) receives the compound ingredients and disperses them to the extent needed to develop target physical properties. The required extent of mixing can be directly related to the ease or difficulty experienced with the same compound in batch mixing. Ingredients such as powders and liquids usually involve no undue problems in accurate continuous metering. The polymer, on the other hand, is often supplied in bales, requiring grinding or pelletizing to provide a form suitable for continuous feeding. Recent demand has caused some polymer producers to offer a range of elastomers in pellets or powder form, thereby improving the option of continuous mixing. The advent of powdered rubber could revolutionize mixing. The advantage is that most or all of the distribution could be accomplished using low power, high intensity blenders. The resultant blend could then be fed to a simple, relatively low horsepower mixer to masticate the compound and render it useful for fabrication. Vulcanizate properties often appear to depend upon a specific level of work input during mixing. In general it is probably not significant whether this work is provided by a sequence of several machines or by a single mixer, although the properties of all polymeric compositions are path-dependent to some degree.

With the advances that have been achieved in raw materials and product handling systems for batch mixers, it is sometimes hard to draw a clear line between batch and continuous systems. In developing the most useful configuration, the following considerations are suggested:

Advantages of continuous mixing
1. Very high product consistency is possible.
2. Automation levels are highest.
3. Minor formulation changes are readily feasible.
4. Minimal floor space and ceiling height are required.

Disadvantages of continuous mixing
1. Supply of raw materials in free-flowing form may be costly.
2. Cleanup from compound to compound may be very extensive.

1.5 DEVELOPMENT OF THE BANBURY MIXER

The Banbury internal mixer was introduced to the rubber industry in 1917; the first mixers were supplied to Goodyear for mixing tire compounds. For many years these mixers bore numerical designations for different sizes. These numbers approximated the quantity of 22in × 60in mills that a specific Banbury mixer could replace. Soon after the first internal mixers were introduced, increases in speed, power and ram pressure made this relationship no longer meaningful.

The earliest mixers had the same basic fundamentals of operation as those which are in service today. A ram is necessary to push raw

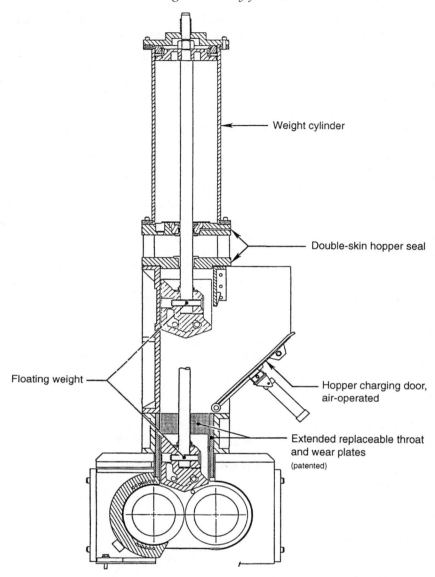

Figure 1.5 An F-series mixer enlarged to accommodate bales and slabs. (Courtesy the Farrel Corporation)

materials into the mixing chamber, two counterrotating rotors perform the mixing action, and a door at the bottom discharges the completed batch to a secondary piece of machinery. Improvements continued to increase the value of this design for mixing applications, despite changes in materials and in expectations.

The F-series Banbury mixers were introduced during the early 1970s. Unlike the earlier mixers, these F-series machines carried designations for each model which defined chamber volume. The F-series Banbury mixers not only introduced many new mechanical improvements, but were designed with the user in mind. Loading and discharge features and maintenance features were designed to emphasize the mixing capabilities of the machine rather than loading, unloading and maintaining.

When the F-series mixer hoppers were enlarged to accept bales and slabs of rubber more readily, the hopper door angle was steepened to assist in more rapid introduction of material to the mixing chamber (Fig. 1.5). The junction between the hopper and the mixing chamber was provided with replaceable throat wear plates. These permit maintenance in an area subject to significant wear because of the action of the ram, well ahead of wear to the chamber. In the latest models, replaceable wear plates have been extended further into the hopper to increase the stiffness of the assembly.

Table 1.2 Batch size and machine volume comparisons for Banbury mixers (Courtesy the Farrel Corporation)

	11D-2W	11D-4W	F-270-2W	F-270-4W	F370	F620
Bore diameter (in)	22.53	22.53	22.56	22.56	27.55	30.03
Gross chamber volume (liter)	436	436	489	489	784	1478
Each rotor volume (liter)	95	108	109	116	185	403
Net chamber volume (liter)	237	220	271	257	414	672
Effective volume reduction weight in throat	3.6	3.6	11.5	11.5	21.0	21.0
Typical batch size at 1.13 sp.g. and 70 Mooney (lb)	460	410	505	480	770	1135
Typical batch size at 1.13 sp.g. and 90 Mooney (lb)	436	388	478	456	730	1070

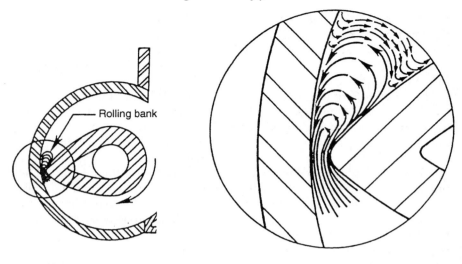

Figure 1.6 Agglomerates are broken down by action of the rotor against the side of the mixing chamber. (Courtesy the Farrel Corporation)

The endframes have been significantly strengthened and the access to the dust seals has been enlarged to assist maintenance or replacement. Within the mixing chamber the rotor endplates have been converted to a one-piece design, eliminating occasional contamination associated with older, two-piece designs. This design also prevents the endplate being dislodged from its fitting, reducing incidence of mechanical failure. Rotor journals are now tapered, eliminating the need for bearing sleeves. Overall manufacturing tolerances have been reduced, lowering vibration and yielding longer useful service life.

The wall thickness of the sides has been increased by nearly a factor of 2 over older models, in response to the higher loads experienced with modern mixing procedures (Chapter 2). The dual-circuit cooling design and the size and location of the bored cooling channels provide highly improved heat transfer. The discharge area has been significantly increased to permit more rapid batch discharge, reducing the interval between batches and adding to productivity. The toggle latch locking mechanism for the drop door has been made more positive in action, but with fewer moving parts, requiring less maintenance.

The most common Banbury sizes are given in Table 1.2. Rotor volumes are measured by immersion displacement, chambers volumes by measurement and calculation. It is entirely feasible to convert an older D-series Banbury to the improved F-series equivalent. In many cases, the existing drive and hopper can be incorporated in the conversion.

As shown in Fig. 1.6, intensive mixing – the breakdown of

agglomerates leading to a high level of dispersion – occurs by the action of the rotor against the side of the mixing chamber. Extensive, or distributive, mixing is accomplished by the continuous moving and shearing of the batch by the rotors; it occurs mainly between the rotors and the ram, and is influenced by the geometry of both. Intensive and extensive mixing must be considered, not only in machine design, but also in regard to mixing procedures.

1.6 OPERATING VARIABLES

The major variables in mixer operation are ram pressure, charging procedure, rotor speed, batch size and coolant temperature. With a properly designed and well-maintained mixer, some selection of these variables will optimize the mixing of almost every rubber compound yet devised.

1.6.1 Ram pressure

The major purpose for application of pressure to the ram is to drive the raw materials into the mixing chamber and to prevent their upward exit

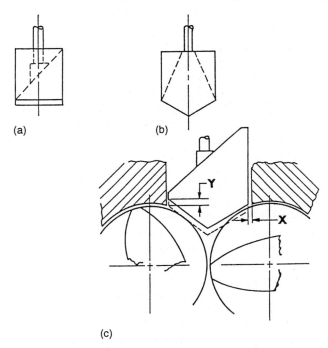

(a)   (b)

(c)

Figure 1.7 Ram configurations: (a) flat bottom, (b) double-gable V-bottom, (c) single-gable V-bottom. Configuration (c) is now considered as the standard. (Courtesy the Farrel Corporation)

during mixing. Increasing ram pressure beyond this point is often found in practice, from a folkloric belief that it will speed the mixing action or otherwise improve it. In fact, it usually has the opposite effect. Too high a pressure can impede rotor action needed for extensive mixing.

The proper seated ram position is shown in Fig. 1.7. The most popular ram configuration, now considered standard, is the single-gable V-bottom (Fig. 1.7 (c)). Somewhat less popular is the double-gable V-bottom (Fig. 1.7 (b)), which facilitates addition of powder or liquid ingredients over the ram while it is in the down, or mixing, position. This procedure is not recommended as it increases wear and maximizes hangup of compound from one batch to the next. It is in exactly the same category as dumping the batch with the ram in the down position. Nevertheless, both procedures retain popularity with those more interested in saving a few seconds of mix time than in machine life or batch-to-batch contamination. The flat-bottomed ram (Fig. 1.7 (a)) may still be found in older machines.

Because of the variation in air cylinder diameters, ram pressure should not be obtained from a pressure gauge on the supply line, but measured directly on the batch. Such data can then be transferred from one mixer to another. High ram pressure is useful in charging the mixing chamber, but a lower pressure is more desirable during the actual mixing. This has led to the development of multiple-ram pressure control systems. Two or three zones are preset with individual regulators to specific pressure levels. This may be automated or controlled by the mixer operator. It can include provision for relief of ram pressure if certain limits on power draw are exceeded, or even if the change in power draw indicates a likelihood that such limits may be reached. This is particularly useful in moderating power surges with upside-down and sandwich mixes (Chapter 2). It also provides the means for staying below certain power plateaus that, in some locations, strongly affect plant utility costs.

The importance of ram pressure and position led to the development of the automatic sensing device shown in Fig. 1.8. By means of its output graph, the ram position indicator can provide a good profile of the mixing action and a guide to proper batch size. Typical profiles are shown in Fig. 1.9. Here three batches of the same compound are mixed to a total elapsed time, T. The desired ram action is shown in Fig. 1.9 (b); there is strong ram up-and-down displacement for the first quarter or third of the total mix time, corresponding to incorporation of ingredients. This is followed by a longer period of reduced ram action, corresponding to distributive and dispersive mixing. Figure 1.9 (a) shows an oversized batch of the same compound. Almost to the end of the batch, the mixer struggles to incorporate the excess of ingredients. Dispersion and uniform distribution will not be optimum if the ram does not

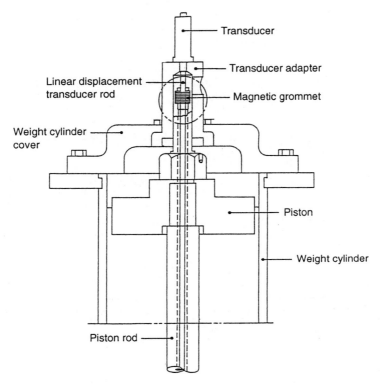

Figure 1.8 The ram position indicator provides a good profile of the mixing action and a guide to the appropriate batch size. (Courtesy the Farrel Corporation)

reach a fully seated position in the first third of the mixing cycle. Figure 1.9 (c) depicts the same batch below optimum batch size; the ram seats almost immediately; distributive and dispersive mixing are strongly impeded.

At optimum batch size, ram displacement charts can be used to fine-tune the ram pressure program so as best to approximate Fig. 1.9 (b). Besides this, batch-to-batch charts can spot equipment malfunction and aid in diagnosing improperly weighed batches.

1.6.2 Rotor speed

Nearly all new Banbury mixers are now equipped with variable-speed rotors, but they are seldom used to vary the rotor speed during the mix (a practice that decreases the useful life of the drive components). Instead

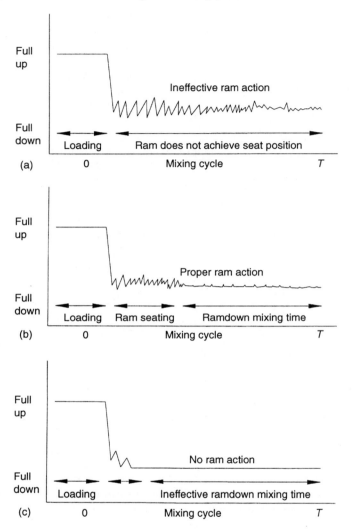

Figure 1.9 Typical mixing profiles: (a) batch oversized, ram does not seat; (b) correct batch size, ram seats in time $0.25T-0.33T$; (c) batch undersized, ram seats immediately. (Courtesy the Farrel Corporation)

the optimum rotor speed for a particular batch is determined. Output, power draw and temperature all rise sharply with increasing rotor speed. For a typical batch, Fig. 1.10 shows temperature versus time at different rotor speeds. Lower rotor speed settings can be used with heat-sensitive compounds, or to prolong the mix when there are severe

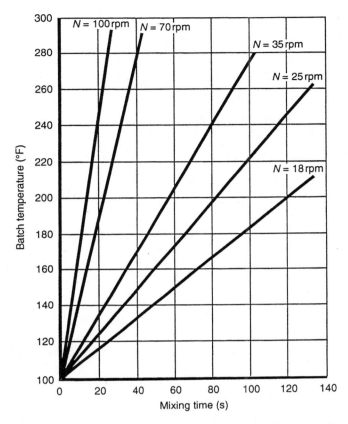

Figure 1.10 Effect of rotor speed on a typical batch: batch temperature plotted against mixing time. (Courtesy the Farrel Corporation)

problems in incorporation or dispersion. Mixes are sometimes run at higher speeds so as to reach maximum output, and they must be checked carefully for uniformity.

Automated control systems have been developed to introduce minor variations in rotor speed during the batch to achieve a specific programmed time–temperature relationship, to compensate for batch-to-batch variation in raw materials.

Rotor condition is also a factor affecting output, temperature and other variables. The normal useful mixer life is gauged by the clearance of the rotor tips and the chamber wall. Throughout this period, the change in clearance is relatively minor (once it becomes less than minor, rebuilding is essential). But during its useful life the geometry of the rotor tip becomes degraded. A rotor cross section is shown in Fig. 1.11. Normal

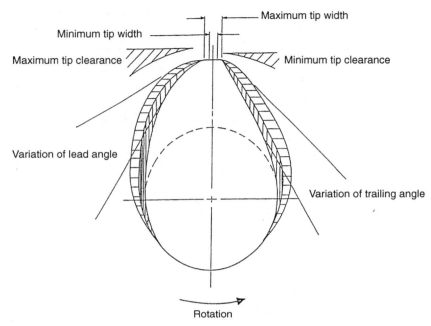

Figure 1.11 Rotor cross section: normal wear leads to a rounding of the rotor tip and may affect the leading and trailing edges. (Courtesy the Farrel Corporation)

wear leads to a rounding of the rotor tip (increased tip width) and may affect the leading and trailing angles. Changes to the leading angle tend to reduce the efficiency with which ingredients are incorporated; changes to the trailing angle tend to reduce the distributive mixing efficiency. Rounding of the rotor tip gradually leads to reduced shear rate and lower dispersion.

Rotor design is also an important consideration. Typical designs are given in Fig. 1.12. and related data in Table 1.3. The length of the wing, or flight, controls internal flow, laterally and from one half of the mixing chamber to the other. It also affects the extent of dispersive mixing carried out with each revolution. The angles of the wings, called **helix angles**, also affect flow and distributive effectiveness. The length-to-diameter ratio of the rotors affects optimum batch size; scale-up from one size of mixer to another can therefore be more complex than a ratio of the mixing chamber volumes.

Conventional Banbury mixers operate with the 'fast' rotor running at higher revolutions per minute (rpm) than the 'slow' rotor, producing a **friction ratio**, as shown in Table 1.3. A new rotor design has

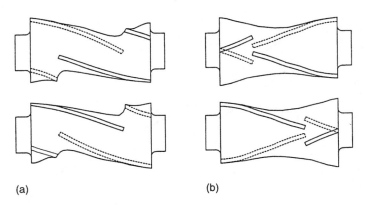

(a)   (b)

Figure 1.12 Rotor designs: (a) Farrel Synchronous Technology (ST) and (b) a standard mixing rotor. (Courtesy the Farrel Corporation)

Table 1.3 Typical rotor dimensions (in inches) for Banbury mixers (Courtesy the Farrel Corporation)

	11D-2W	*11D-4W*	*F-270-2W*	*F-270-4W*	*F370*	*F620*
Tip/wall clearance	0.340	0.340	0.342	0.342	0.375	0.393
Rotor flight length (long/short)	22.5/13.5	21.3/11	24.6/14.8	23.6/11.5	25.2/12.3	38.0/22.1
		20.5/10		23.0/11	23.5/11.5	36.5/20.5
Rotor diameter	21.88	21.88	21.90	21.90	26.79	29.27
Body length	31.88	31.88	34.88	34.88	37.24	59.81
Friction ratio	1.13	1.13	1.13	1.13	1.13	1.18

recently been introduced (Fig. 1.12), called Synchronous Technology (ST) rotors, which run at even speed. Synchronous Technology and ST are trademarks of the Farrel Corporation. This innovation is discussed in Chapter 14.

1.6.3 Batch size

Optimum batch size is calculated by multiplying the volume of the mixer (kilograms or pounds of water at specific gravity = 1.0) by the specific gravity of the compound, and by the **fill factor**. The problem has always been the estimation of the fill factor, typically based on experience and

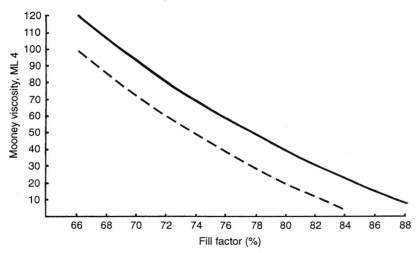

Figure 1.13 How viscosity affects batch size; batch size = compound density × mixer volume × fill factor: (solid line) masterbatch at typical mixer speeds, 60 psi ram pressure on batch; (dashed line) final mix at typical mixer speeds, 40 psi ram pressure on batch. (Courtesy the Farrel Corporation)

observation. Batch size should obviously be related to compound viscosity; soft, oil-extended compounds should have a higher fill factor than hard, high viscosity stocks. A good starting point is given in Fig. 1.13, where suggested fill factor is plotted versus Mooney viscosity (100 °C, ML 1 + 4) for first- and second-pass batches at common ram pressure settings. Special-purpose polymers that are notorious for requiring high batch sizes, such as butyl rubber, will shift the curves upwards by 1–2%. Those famous for low fill factors, such as neoprene, may be shifted down a like amount. (Chapter 5 gives #11D batch sizes for a variety of compounds; these may be used to make special case adjustments to Fig. 1.13.)

1.6.4 Coolant temperature

Most internal mixers once used refrigerated (or otherwise cooled) water as the coolant for the sides, rotors and door, but this has radically changed with the advent of drilled sidewalls having efficient heat transfer characteristics. Chilled water is actually counterproductive when used with the drilled sidewalls characteristic of all modern mixers. Overchilling leads to condensation of moisture on the walls of the mixing chamber, reducing wetting of the metal by the compound (essentially reducing its coefficient of friction).

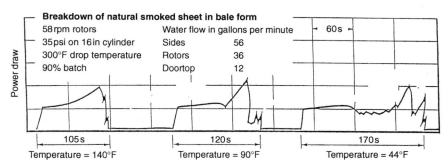

Figure 1.14 How coolant temperature affects power draw: data for a Farrel 3D Banbury with four-wing rotors. (Courtesy the Farrel Corporation)

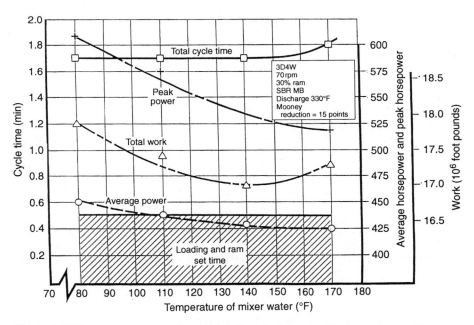

Figure 1.15 Temperature control systems allow independent temperature settings for sidewalls, rotors and door. (Courtesy the Farrel Corporation)

Table 1.4 Flow rates for Banbury mixers in gallons per minute (Courtesy the Farrel Corporation)

	1	3D	9/9D	F80	F270/11D	F370	F620/27D
Sides	30	60	60	120	120	140	180
Rotors	20	35	40	40	60	100	150
Doortop	10	10	10	10	15	30	30

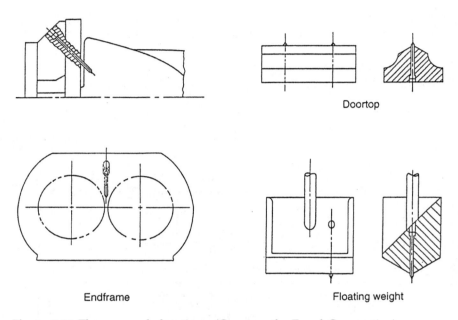

Doortop

Endframe

Floating weight

Figure 1.16 Thermocouple locations. (Courtesy the Farrel Corporation)

Instead, use of controlled-temperature water as the coolant provides sufficient transfer to remove the shear-generated heat of mixing, but maintains stronger wetting of the chamber walls (essentially higher coefficient of friction), leading to faster and more effective mixes. An example, the mastication of smoked sheet natural rubber, is shown in Fig. 1.14. Power draw versus mix time is recorded for controlled water settings of 44, 90 and 140 °F, using a 3D Banbury. All of these yielded approximately the same Mooney viscosity. The improvement in cycle time with warmer coolant settings is very significant.

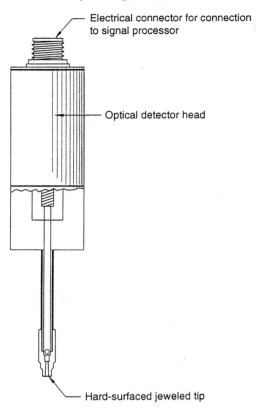

Electrical connector for connection to signal processor

Optical detector head

Hard-surfaced jeweled tip

Figure 1.17 This infrared thermometer has a jeweled tip to prevent it from becoming damaged. (Courtesy the Farrel Corporation)

Figure 1.15 shows several parameters reported during the mixing of SBR black masterbatch to a consistent reduction of 15 Mooney viscosity points at various temperature settings. Higher temperatures noticeably lower the average and peak power, and, up to a point, the total work input too. The temperature settings in these examples are for all three zones: sidewalls, rotors and door. Most of the cooling is through the sides of the Banbury. With compounds that are sticky or tend to hang up, the rotors and door are often set somewhat lower than the body. The water flow rates that can be reached in these zones of common mixers are given in Table 1.4.

Equally important is the temperature of the batch itself. In older Banbury mixers, the thermocouple(s) are located in one or both sidewalls, since sliding doors prevented the use of doortop locations. In more modern machines, doortop (and occasionally ram weight) locations are

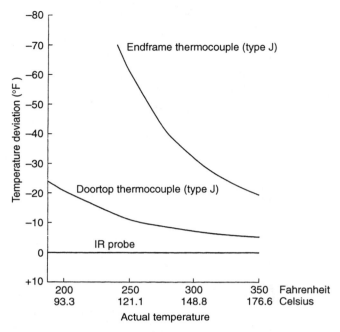

Figure 1.18 The infrared thermometer shows dramatically smaller deviation than the two type J thermocouples. (Courtesy the Farrel Corporation)

much more widespread (Fig. 1.16). The doortop position is generally the most practical. Standard installation uses type J thermocouples. In laboratory tests these thermocouples can register a 100 °C temperature rise in about 2 s. But a batch of rubber being mixed is far from an ideal environment. In practice the response is affected by rotor speed, batch size, compound viscosity and thermal conductivity. This has led to the development by Farrel of an infrared thermometer with a jeweled tip to prevent damage (Fig. 1.17). The infrared signal is transmitted through this window to the detector via an optical fiber bundle. Figure 1.18 compares the deviation from the temperature measured immediately after dropping the batch with the deviation obtained using sidewall and doortop thermocouples. The superiority of the infrared method is dramatic. The development of accurate batch temperature measurement capability in production-scale equipment makes it possible to improve automated control of internal mixing.

There is no question that innovation in the refinement of mixing machinery has responded to, and kept pace with, developments in material science that have broadened the application of elastomeric compositions.

2

Mixing cycles and procedures

Richard F. Grossman

2.1 COST OF INTERNAL MIXING

The goal in mixing is to provide compositions having useful properties and suitable processability with as high a level of consistency as possible. The terms *useful* and *suitable* are determined by the application; what is suitable for a sink stopper might not be suitable for an O-ring. In almost all applications there are criteria for the attributes that characterize whether a composition is suitably mixed; the criteria may vary, but they exist nonetheless. And it is almost always the case that these criteria must be met with optimum efficiency; that is, with the maximum output per expenditure of capital and energy.

At the present time, a first approximation is that a general-purpose rubber compound will cost about one cent per pound per minute in the internal mixer; covering energy, labor, stack losses and disposal costs. With very large mixers, running costs may be slightly lower; with specialty compounds, somewhat higher. Fully overheaded costs (including depreciation) may approach two cents per pound per minute. A highly automated system with computerized controls, automatic weighing and delivery, will draw 3–4 cents per pound per minute. The same numbers apply if one uses a custom compounder to do the mixing, instead of carrying it out in-house, assuming the compounder is sagacious enough to expect net profits to exceed investing in savings bonds. As a result, efforts spent to design mix cycles that are both effective and efficient are very worthwhile. The objectives when mixing rubber compound are summarized in Table 2.1.

The Mixing of Rubber
Edited by Richard F. Grossman
Published in 1997 by Chapman & Hall, London. ISBN 0 412 80490 5.

Table 2.1 Objectives when mixing rubber compound

● To provide compositions having suitable (which usually means marketable) properties, processability and cure characteristics that are as consistent as possible

● To do so with maximum output per expenditure of energy and capital

Table 2.2 Steps in the mixing process

● Wetting of the ingredients
● Distribution of ingredients
● Dispersion of agglomerates
● Reduction of viscosity
● Reactions and interactions

2.2 UNIT OPERATIONS IN MIXING

As a simplification, mixing is customarily divided into the operations of wetting (or incorporation); distribution of ingredients to a more or less visually homogeneous state; dispersion, the reduction of agglomerates into smaller particles; reduction of compound viscosity; plus any chemical reactions or heat interactions that may result from mixing (Table 2.2). This is not to say that the actual mechanics of mixing may be so cleanly divided; it is our picture of the mixing process that has been compartmentalized. But the division of mixing into these unit operations is a useful construct.

Incorporation and distribution are required in all mixes. (When these processes do not occur to a sufficient extent, the batch is considered unmixed.) Some batches, particularly those containing reinforcing carbon black, are also characterized by substantial dispersive mixing, necessary for development of optimum physical properties. Dispersion will not reach effective levels if it has not been preceded (or accompanied) by incorporation and distribution. Unmixed pockets of carbon black usually result not from inherent problems with agglomerates, but from poor wetting and consequent inadequate distribution. Some dispersion, e.g. breakup of pelletized fillers, occurs prior to and along with incorporation. Thus, a problem with dispersion can sometimes interfere with incorporation and distribution, although the reverse is more common. The factors affecting incorporation are listed in Table 2.3.

Internal mixing invariably lowers the average molecular weight and therefore the viscosity of the base polymer, through heat and shear,

Table 2.3 Factors affecting wetting of ingredients

- Interaction parameters
- Flow parameters
- Surface area
- Time, temperature and pressure

Table 2.4 Solubility parameters of common ingredients[a]

	Hildebrands	MPa$^{1/2}$
FKM, silicones	7–7.5	14–15
PE, EPM, EPDM	8	16
NR, BR, IIR	8–8.5	16–17
SBR, low MW resins	8.5–9	17–18
CR, CSM, some NBR	9–9.5	18–19
Typical lubricants + process aids	8.5–9.5	17–19
PVC, ECO, ACM, some NBR	9.5–10	19–20
Carbon blacks	12–15	24–30
Clay, whiting, talc	13–14	26–28
Silicas	14–18	28–36

[a]To convert from hildebrands to MPa$^{1/2}$, multiply by 2.05.

unless specific chemical reactions are introduced to extend molecular size. This does not guarantee that increased mixing will lower the compound viscosity; it depends on whether the loss of molecular weight exceeds any interaction with polar fillers. Consistent processability for article fabrication therefore depends on rigorous use of the same mixing procedure. Batches that are remixed (e.g. to correct perceived errors in composition) may or may not show processability that has been improved, but invariably they show some alteration to their process characteristics.

Usually the most significant factor affecting wetting is the polarity of the ingredients. An approximation is given by the solubility parameter, and typical values are listed in Table 2.4. (Estimates of solubility parameter are given both in hildebrands, the unit found most commonly in the literature, and also in SI units, where joules in place of calories leads to the square root of megapascals.) Ease of wetting increases as the difference between solubility parameters decreases. In the case of two polymers mutually wetting, differences greater than 1 hildebrand can cause problems in mixing. The solubility parameters (actually, interaction parameters) of fillers typically are much higher than the solubility parameter of the polymer. In this case, incorporation occurs as the polymer flows around the particle, provided the polymer is able to wet itself and reform after surrounding the solid. A polymer whose surface tends to be less polar than the bulk of the phase (polyethylene, butyl rubber) may, in some compounds, have problems with self-adhesion during filler incorporation. More prevalent is the case where remnants of a partitioning agent (used to keep bales apart or powdered polymer from massing) interferes with the polymer wetting itself. Mixing techniques have been devised to cope with these situations.

Speed of incorporation also depends on polymer flow; generally it decreases with increasing viscosity. But distribution and dispersion may be assisted by increased polymer viscosity, assisted to a greater extent than wetting is hindered. In general, as the surface area of the polymer is increased, e.g. by grinding it to a powder or converting it to pellets before mixing, wetting of other ingredients is favored. Sometimes this is a very useful technique. Once again, its effectiveness depends on rapid rewetting after incorporation of other ingredients.

Wetting of ingredients is also greatly assisted by pressure and temperature (within certain limits) and, of course, by increased mixing time. The limiting factor on the substitution of rotor speed, pressure and batch temperature for mixing time – the most expensive input – is the molecular weight (MW) of the polymer. If the MW is low enough for Newtonian viscosity to apply (e.g. low-MW PE), the intensity of mixing can be increased almost to the onset of cavitation without inducing substantial degradation, as in the high speed mixing of plastic color concentrates. With typical plastics and elastomers, useful properties depend on the MW of the polymer, which can be sacrificed only to a minor extent in the interest of output. The thrust of development has attempted to achieve the best compromise in mixing procedures.

Table 2.5 lists a 'typical' rubber formulation (not invented, but taken from a recipe for a belt cover). The example includes three grades of rubber, two of carbon black, two types of clay, two antioxidants, a tackifying agent, a release agent, two process oils and a complex cure system. (Historically, the limit to the number of ingredients has been the number of lines available on the mix card.) Possibly unwittingly, the formulator has made great use of solubility and interaction parameters. The mixture of the two carbon blacks has a broader particle size distribution than either of the individual grades; this factor will increase the speed of incorporation. The blend of two process oils has broadened the solubility parameter, facilitating incorporation. This is also true of the polymer blend if the components are of closely matched solubility para-

Table 2.5 Typical rubber formulation

Rubber no. 1	165 lb	6 oz
Rubber no. 2	39	8
Rubber no. 3	17	8
SRF black	45	
FEF black	39	8
Hard clay	42	
Soft clay	17	
Zinc oxide	11	2
Stearic acid	2	4
Antioxidant no. 1	2	3
Antioxidant no. 2	3	4
Tackifying resin	8	10
Release agent	1	8
Process oil no. 1	26	
Process oil no. 2	14	8
Process aid no. 1	2	4
Process aid no. 2	3	2
Sulfur	4	4
Primary accelerator	3	8
Secondary accelerator	1	2
Activator	0	10
Retarder	0	8

meter. The blend of mineral fillers will often incorporate more rapidly than either of the components. This does not mean that every recipe requires 20+ ingredients, or that every blend is an improvement over a single ingredient.

The method of attacking problems in formulation by using blends of all kinds, although it commonly assists ingredient incorporation, does introduce other problems. In addition to maximizing the chances for weighing errors, the length of the recipe makes it more difficult to decide on the order of addition for the ingredients. The compounder generally follows the first law of mixing: Put ingredients that seem compatible (have similar solubility parameters) together; keep those that don't apart. As the recipe becomes longer, it is less likely that the technologist will be able to guess the relative polarity of ingredients. Assistance by use of solubility parameters may be helped by several sources [1–3].

Table 2.6 Factors affecting distribution

- Polymer viscosity
- Shear rate
- Dependence on wetting

The factors affecting distribution (Table 2.6) include polymer viscosity, shear rate and the speed of wetting. Distribution into a visibly homogeneous phase cannot occur more rapidly than the wetting of ingredients. Shear rate (and therefore rotor speed, power input and pressure) must usually be matched to polymer viscosity and rate of wetting in order to maximize output with the most consistent product. Almost all the practical questions about mixing center on the best way to achieve such a match.

2.3 SINGLE-PASS VERSUS MULTIPLE-PASS MIXING

The first question is often, How many passes should be used to mix a given compound? Single-pass mixing predominates in the United States; two-pass mixing is much more common in Western Europe. There are several reasons for using more than a single pass (Table 2.7): if substantial dispersive work is needed, temperatures may well exceed the safe limits at which curatives can be added. Alternatively, further processing may require a cure system too active to be added to the first pass, even if intense dispersion is not required. And there are numerous cases where unaccelerated compound is preferred, for reasons of shipment or storage, or to facilitate screening out of discrete impurities in critical applications (roll covers, wire insulation, printing blankets, etc.) Furthermore, it may be desired to use one masterbatch with several cure systems, or to blend several to achieve a final compound. This is common in cases where combinations of marginally compatible ingredients are needed in the same recipe, and is an established method of dealing with certain wetting problems. Finally, very high molecular weight polymers

Table 2.7 Two-pass mixes

- More than just wetting and distribution needed
- Process requires fast cure activity
- First pass must be screened, stored or transported
- First pass is used in several final compounds
- Blends of several first-pass mixes used in compound

Table 2.8 Typical SBR formulation

SBR 1606	162 lb
Hard clay	50
Whiting	50
Factice	10
Zinc oxide	3
Stearic acid	1
Antioxidant	1
Process aid	2
Process oil	30
Sulfur	2
CBS	1.2
MBTS	0.5
DPG	0.5

are sometimes mixed in multiple passes to lower viscosity and promote further processing. The products of such operations may provide a better balance of properties than simply using a lower viscosity grade of the same polymer (but probably not as often as thought; this should be checked by relevant experiments).

An example of a compound that might be mixed either in two passes or in a single pass is given in Table 2.8. If used to produce high quality sink stoppers or force cups, this compound would no doubt be mixed in a single pass. The safe cure system and the fact that the carbon black is already dispersed in the SBR 1606 enable efficient single-pass mixing. On the other hand, if the compound were to be used to fabricate typewriter platens, it would be advisable to screen impurities from a first pass, then to accelerate in the second. In a well-maintained internal mixer of recent design, the single-pass mix, or mixing the first pass, would probably take 2–4 min, depending mainly on the efficiency of handling compound on exit from the mixer. An additional pass for adding curatives would add another 2–3 min. This cost should be balanced against the cost of reformulation with more expensive ingredients (less likely to contain screenable impurities), and cleaner process conditions.

An interesting example of a compound mixed in two passes is given in Table 2.9. It would be difficult to achieve useful carbon black dispersion in a single pass without reaching temperatures that would scorch the compound (140 °C). In practice the first pass is mixed with 80–100 phr black with proportional increase in antioxidant. In the second

Table 2.9 XLPE low voltage insulation

Low density PE	100 lb
Antioxidant	0.3
MT black	40
Dicumyl peroxide	2.6

Table 2.10 Butyl insulation

Butyl, low unsaturation	100 lb
Calcined clay	100
Platy talc	50
Zinc oxide	5
Stearic acid	1
Paraffin wax	2
Quinone dioxime (QDO), 70%	3
MBTS	4
Red lead, 90%	6

pass, the cross-linking agent and additional polymer are added to reach the desired overall recipe. In this way only 40–50% of the compound is mixed twice, lowering the process costs. This technique is attractive with polymer-rich compounds. It may often be advisable to use different viscosity grades of the same polymer in the two passes, because the viscosity reduction is greater in the part that is mixed twice; but this is not always needed.

Table 2.10 gives an example of a two-pass mix where typical manufacturers would add only curatives in the second pass (percentage notations are listed in the case of hazardous ingredients, representing the activity of the pure material in the dispersions most commonly used). The cure system is much too active for single-pass mixing, the compound (wire insulation) requires screening, and processability (extrusion) is much improved by a two-pass mix. If filler extension were increased to bring the polymer level to 25–30% and safe cures substituted, e.g. in butyl sheeting, then single-pass mixing would be used.

2.4 TYPES OF MIX CYCLE

The general types of cycles used with internal mixers are often divided into the categories listed in Table 2.11 [4]. The designation 'conventional

Table 2.11 Types of mix cycles

- Conventional
- Late oil addition
- Upside down
- Sandwich mixes

Table 2.12 Heater cord insulation

Polychloroprene	100 lb
Hard clay	70
Whiting	55
Zinc oxide	5
Magnesium oxide	4
SRF black	4
Antioxidant	3
Process oil	12
Protective wax	3
MBTS	1
ETU, 95%	1

mix' is usually taken to mean that all the polymer is added first then massed in the mixer. This is followed by incremental addition of fillers and liquid ingredients, starting with those most difficult to incorporate and distribute, and those which require dispersive work. Easy-to-incorporate ingredients and curatives are added later in the cycle. This type of cycle, a relatively slow mix, is very suitable for older internal mixers having comparatively low shear rates and modest cooling capability. This is not to imply suitability for poorly maintained older equipment; no procedure will produce satisfactory results under such conditions. With well-maintained, properly reconditioned internal mixers, built 20 even 50 years ago, conventional mix cycles will yield generally suitable mixed compounds.

Another common reason for using the conventional mix cycle is to provide an intentionally slow cycle. This may become desirable when a compound that was originally two-pass mixed, as in Table 2.12, is condensed to a single pass for economic considerations, despite an active cure system. The subject compound is an integral wire covering for household or light industrial use. Thorough incorporation and distribution are essential to prevent electrical failure; a slow conventional mix is generally used.

A conventional mix cycle does not imply absence of innovation. Consider, for example, the compound given in Table 2.13, a common natural rubber/SBR blend. The first mix procedure is a standard conventional mix, as might have been recommended in a prewar textbook of rubber technology (except for the use of SBR). But the second procedure will also run well in an older (but well-maintained) mixer at greatly improved output. There is a further consideration: blends of clear polymer and black masterbatch (SBR 1605) can pose incorporation problems. Differences in viscosity, partitioning agents used to keep bales from sticking, and damp fillers are potential causes of wetting problems. The most common symptom is small particles of unmixed clear rubber in a compound matrix. Certainly such incorporation problems can be attacked through compounding, and improved through selection of grades to be blended. Nonetheless, a good general approach is to convert the clear rubber to compound in the mixer, then to add black

Table 2.13 Natural rubber/SBR blend

Ingredients	
SBR 1605	180 lb
SMR5	100
Mineral filler	100
Small ingredients	21
Process oil	24
Curatives	8

Conventional mix		
0 min		Load all rubber
2 min		Load 1/2 filler, 1/2 oil
3.5 min		Load rest except cures
~5 min	165 °F	Load cures
9.5 min	220 °F	Dump

Revised mix		
0 min		Load NR; load filler, oil
		Load 90 lb SBR 1605
1 min		Load rest; load 90 lb SBR
~5.5 min	215 °F	Dump

(or other) masterbatch. What is being done is effectively to broaden the solubility parameter of the clear polymer, giving it greater compatibility with another, previously compounded elastomer. Improved compatibility underlies the increased output of the revised mix procedure of the example of Table 2.13.

Logic in the order of addition is a more potent tool than the typical remedy for poor incorporation: mix longer.

2.4.1 Late oil addition

Another method with a long history involves late oil addition. Suppose, for example, the compound of Table 2.13 was being mixed not with SBR black masterbatch but with natural rubber, clear SBR and carbon black. For safety, many common cure systems would have to be omitted and the batch mixed in two passes. Assume now that the rubber technologist has at his or her disposal not a well-maintained mixer purchased in 1958, but the most up-to-date equipment. A useful technique would be to add all ingredients except oil (and, of course, curatives) at time zero and to mix with sufficient power input to incorporate, distribute and disperse within 1–2 min. At this point the batch temperature would probably have reached 130–140 °C; oil addition would then begin at a rate selected to maintain but slightly moderate the mixing action.

The technique of late oil addition was once widely used in first-pass mixing of tire components to yield a combination of high output and good black dispersion (at the expense of heavy power input). It is currently less prevalent because of the availability of high structure carbon blacks that can be dispersed even in the presence of process oil, and because of increasingly sensitive temperature control available with modern intensive mixers.

When using high viscosity oils, high output is possible only if the oil is preheated to allow rapid addition. If heavy oil is not able to be preheated, its rate of incorporation is greatly affected, perhaps limiting the site to a conventional mix cycle. (This can be compensated, at considerable expense, by use of process oils predispersed on powder carriers. The approach is common when a compound is mixed only occasionally.)

Another area where viscous oils and late oil addition cycles are popular is in the mixing of hard rubber compounds. A typical example (a floor caster) is given in Table 2.14. Here the combination of hard rubber dust and heavy aromatic oil, each individually a potential problem in incorporation, together serve to bring the batch together rapidly at cool temperatures, always desirable with high sulfur loadings. (If a batch with 20 phr sulfur in an unsaturated elastomer were to reach 140–150 °C, say through thermocouple failure or operator oversight, ignition is quite likely.)

Table 2.14 Late oil addition with hard rubber compounds

Ingredients	
SBR 1502	100 lb
Hard rubber dust	125
Whiting	150
Sulfur	20
Small ingredients	28
Aromatic oil, heavy	32

Modified late oil addition		
0 min		Load whiting; load rubber, sulfur, small ingredients
1 min		Load rubber dust and oil
5–6 min	215 °F	Dump

Table 2.15 Mixing with powdered masterbatch

SBR 3650 powder	280 lb
Hard clay	85
Whiting	25
Hydrocarbon resin	8
Zinc oxide	7
Stearic acid	2
Sulfur	3.5
Accelerators	2.3
Naphthenic oil	40

Although the volume of hard rubber compounds has been greatly reduced, late oil addition has gained in popularity through application in other areas. It is very useful in mixing compounds with powdered elastomers; an example is given in Table 2.15. Here powdered black masterbatch (SBR 3650) and all other powder ingredients are added to the mixer, which is then run as a blender for 20–60 s, typically without applied pressure. Process oil is then added rapidly and the ram lowered onto the batch. A quick surge of power completes mixing extremely rapidly. To be successful, the equipment must be able to accommodate substantial power surges as the batch takes hold. Overall wear and tear

Table 2.16 Rubber/plastic blend, late oil addition

PVC/VA copolymer	60 lb
CPE, elastomeric type	40
Hydrated alumina	88
Whiting	32
Antimony oxide	7
Lead stabilizer, 90%	7
Antioxidant	1
Chlorinated paraffin	30
Liquid epoxy resin	3
Trimethacrylate monomer	5

on the mixer is lower than mixing with clear rubber and carbon black, compensating for slightly higher material cost.

The most significant area using late 'oil' addition is in mixing plasticized PVC, CPE and their blends with powdered elastomers. A typical plasticized PVC/CPE blend is given in Table 2.16. This compound is usually mixed in three stages: (1) addition of all powders to the mixers, (2) blending, (3) addition of all fluids (chlorinated paraffin, liquid epoxy resin and methacrylate monomer). A frequent variation is to premix all ingredients, powder and fluids, in a blender. The dry blend is then metered to a batch mixer or mixing extruder. Capable of extremely high output, this variation is useful in cases where the powdered polymer can absorb blended fluids and remain more or less a powdered solid. For the compound of Table 2.16 a paste-like semisolid would result instead, limiting high speed mixing to the late oil addition sequence suggested above.

2.4.2 Upside-down mixing

Conventional and late oil addition cycles are most often used with polymers having good self-adhesion, i.e. polymers that wet themselves easily. Upside-down mixing is often used with polymers having limited self-adhesion, such as EPM and EPDM. It is also used frequently when the, percentage of polymer is low (15–25%); this is an extension of the limited self-adhesion case. In a conventional mix, low polymer content is difficult to mass at the start of the cycle simply because so little is present. If the polymer content is low, even a rubber with strong self-adhesion (e.g. polychloroprene) may be mixed upside down.

Table 2.17 Upside-down mix cycle

Ingredients	
EPDM polymer(s)	125 lb
SRF black	125
Soft clay	160
Small ingredients	12
Process oil	80

		Cycle
0 min		Load black; start oil; load clay and small ingredients
		Hold 30 s
		Load rubber; ram down
4–6 min	130–140 °C	Dump

A typical upside-down order of addition is shown in Table 2.17. As given, the compound is (most likely) first-pass without cures. But EPDM compounds such as this are most often single-pass mixed with the upside-down technique. Final dump temperatures are usually lower than with other procedures, thus encouraging single-pass mixing.

It is important to note that strong dispersive work on reinforcing fillers cannot be carried out using this procedure. If fine particle black (e.g. HAF) is present, the compound should be mixed conventionally or with late oil addition. Most EPDM recipes involve little filler dispersion, mostly incorporation and distribution. These functions are carried out very efficiently with upside-down mixing. As there is typically a strong power surge when the batch comes together, the technique is most appropriate for late-model internal mixers. This is not to say that older, well-maintained machines cannot run upside-down cycles. If the batch contains substantial oil, older mixers that tend to leak (sliding door closures) are unhelpful.

Another useful area is where the polymer, although having good self-adhesion, is supplied in a form that inhibits self-wetting. An example is chlorosulfonated polyethylene (CSM), where small chips are protected from massing by an effective partitioning agent (Table 2.18). This is no problem in polymer-rich recipes and conventional mixes are easily run. Similarly, two-pass mixes with active cures are readily mixed with conventional cycles. Single-pass mixes with substantial quantities of process oil or plasticizer are usually mixed upside down. As with EPDM, filler, oil and small ingredients are loaded, and the batch allowed to

Table 2.18 CSM cable jacket compound

Ingredients	
CSM, high MW	100 lb
Hard clay	85
SRF black	30
Aromatic oil	50
Small ingredients	8
Curatives	3

	Cycle	
0 min		Load black; start oil; load clay and small ingredients
		Hold 30 s
		Load rubber; ram down
2–3 min	70–80 °C	Load curatives
	100–110 °C	Dump

tumble-blend for a few seconds. Then the polymer is added and the ram lowered, causing a strong power surge; the accompanying shear carries out incorporation and distribution. Curatives may be added with the small ingredients, or later in the cycle if they are strongly active. Alternatively, curatives that are difficult to incorporate may be added with the small ingredients, and the balance added towards the end of the mix.

If little dispersive work is required, most batches of all types can be run upside down. The technique is particularly favored by mixers having four-wing rotors. Even polymers that tend to resist upside-down incorporation (typically soft elastomers in bale form, such as butyl and polybutadiene) may be mixed upside down with little difficulty using four-wing rotors. As heat buildup is greater than with two-wing rotors, minimization by the upside-down procedure is particularly valuable.

The upside-down technique is also very common when the polymer consists of bales of SBR (or other) black masterbatch. Since carbon black dispersion has already been taken care of, it is a natural step to adopt single-pass mixing of the final compound with a high output, cool-running procedure. A typical example is given in Table 2.19. This type of cycle maximizes the value of black masterbatch (as compared to mixing pure polymer with carbon black in a first-pass mix). Blends of the SBR 1600 series may be run easily, also blends of the SBR 1800 series. But SBR 1600/1800, or other blends of masterbatch with greatly different oil levels, are rarely mixed with a simple upside-down procedure. As

Table 2.19 SBR black masterbatch, upside-down mix

Ingredients	
SBR 1805	270 lb
Ppt. silica	75
Chlorinated paraffin wax	100
Antimony oxide	15
Small ingredients	20
Curatives	8

	Cycle
0 min	Load all powders; sweep down
	Load rubber
200–220 °F	Dump

with all mixing techniques, different elastomers, or different grades of the same elastomer, should be added simultaneously only when they are so compatible as to be almost indistinguishable. When this is not the case, blends are best accommodated with a sandwich mix.

2.4.3 Sandwich mixes

In this procedure one polymer (or a close blend) is loaded, and filler plus oil is added. The assumption is that converting the polymer to compound will broaden its solubility parameter. Then the second polymer is added on top, completing the 'sandwich'. Physical form is a common reason for less than complete compatibility (as regards incorporation) in two grades of the same elastomer. This situation is presented in Table 2.20: a blend of a soft, amorphous EPDM supplied in bales and a hard, semicrystalline grade supplied in pellets or in friable bales. If they are simply combined, either in a conventional or upside-down mix, often one (or both) will yield small spots of unmixed polymer. (This is not to say it will always occur with a given blend, or that it does not depend on the rest of the compound, merely that the effect is common.)

In this example the soft, amorphous EPDM is loaded on the bottom, and the remaining ingredients are added and mixed, but the batch is not brought completely together. The ram is then lifted and the partially mixed batch is allowed to expand slightly under the reduced pressure. The semicrystalline EPDM is added, as in an upside-down mix, with a

Table 2.20 EPDM blend, sandwich mix

Ingredients	
EPDM, amorphous, 40 Mooney	75 lb
EPDM, semicrystalline, 80 Mooney	55
Soft clay	300
Process oil	110
Small ingredients	10

Cycle		
0 min		Load amorphous EPDM; load 100 lb clay; start oil; load rest of powder; drop ram
1.5 min		Hold 30 s with ram up
2 min		Load semicrystalline EPDM; drop ram
4–6 min	100–110 °C	Dump

similar power surge as the ram is reapplied. This technique is very effective with such blends, providing an optimum batch size has been chosen. (Of the various types of mix cycle, the conventional procedure provides the greatest forgiveness of deviation from optimum batch size, the sandwich mix probably the least.)

The sandwich mix should also be used with blends of elastomers having different solubility parameters. A common example is the blending of SBR and polychloroprene. Polychloroprene is conveniently loaded on the bottom, fillers and oil are added, and SBR, particularly black masterbatch, placed on top. It is not always clear which elastomer should be the first ingredient, and which the last ingredient. (Sometimes either will do.) The choice for the 'bottom' of the sandwich should be a polymer that masses easily, one that responds readily to a conventional mix cycle. The elastomer used to close the sandwich is typically an elastomer that is useful in upside-down mixing.

When blends of polymers having different polarity are mixed, consideration must be given to how other ingredients will distribute. Polar materials, e.g. carbon black, will prefer association with the more polar polymer. Thus if a blend of elastomer X and elastomer Y with carbon black is mixed by the sandwich technique, the choice of which elastomer is added first will strongly influence the properties of the compound.

Table 2.21 EPDM/LDPE blend, sandwich mix

Ingredients	
EPDM, amorphous, 60 Mooney	45 lb
LDPE, 2.0 melt index	55
Calcined clay	85
SRF black	10
Zinc oxide	5
Antioxidant	1
Paraffin oil	15
Dicumyl peroxide	4

	Cycle	
0 min		Load PE, 2/3 filler, small items
1 min		Load rest of filler and oil
		Load EPDM
3–3.5 min		Load peroxide
5–6 min	100–110 °C	Dump

Preblending X with Y, then mixing conventionally will generate a third set of properties, again probably different. Sometimes the rubber technologist may be able to predict which procedure might be most appropriate for the application in view; sometimes relevant experimentation will be needed. Very seldom could the mixing procedure oscillate between the various orders of addition without substantially affecting the observed properties.

A broad area of application for the sandwich mix is in rubber/plastic blends. An example is given in Table 2.21: a blend of EPDM with low density polyethylene (LDPE). Here the sandwich mix is appropriate for single-pass mixing of a blend that is thermodynamically very compatible, but where reaching equilibrium (incorporation) is hindered by the difference in physical form. Almost any order of addition may be run using a two-pass mix, as long as there is no curative in the first pass that would place a limit on the temperature.

Another common example is PVC and NBR, where the NBR is used not in powder form, but in bales. In this case the PVC would be loaded with any other powders, and all liquids (e.g. plasticizer) would be added. After a brief blending period, NBR is loaded to complete the sandwich. The mix cycle is somewhat slower than late oil addition to a blend of PVC and powdered NBR, but it permits the use of lower priced NBR bales.

2.5 ANALYSIS OF CHANGES TO THE MIX PROCEDURE

Two types of changes to the mix procedure are typically contemplated: revisions that will make one or more of the steps more effective (e.g. incorporation, dispersion), without much loss in efficiency; and revisions that will be more efficient (lead to higher output) without sacrificing effectiveness. The first type usually implies changing the order in which the ingredients are added, i.e. varying the type of mix cycle (as discussed earlier). The elements needed for success are a thorough knowledge of the characteristics of the specific mixing equipment, and an empirical knowledge of the behavior of raw materials, perhaps guided by theoretical considerations.

With regard to improving efficiency, deductions can be made while observing the current mix cycle. In particular, observe the batch temperature and the power drawn by the mixer throughout the cycle. This is

Table 2.22 SBR/polyisoprene blend

Ingredients	
SBR 1605	180 lb
Polyisoprene (IR)	75
N347 HAF	100
Zinc oxide	11
Octylated diphenylamine (ODP)	3
Stearic acid	3
Process oil	25
Batch weight	397

Cycle	
0 min	Load IR; load black; load 90 lb 1605
0.5 min	Loading complete, ram down
1.5 min	Load zinc oxide, 90 lb SBR 1605
3 min	Load oil, stearic acid, ODP
5 min	Dump; hold 30 s; close + load

With #11D type mixer, output = ~ 4000 lb h^{-1}

Average power draw = ~ 250 kW

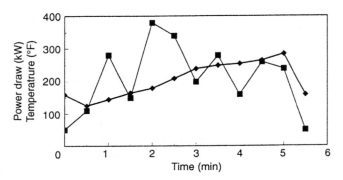

Figure 2.1 SBR/polyisoprene blend, ram down at 0.5 min, second addition at 1.5 min, third addition at 3 min, sweep at 4 min, dump at 5 min: (♦) batch temperature and (■) power draw versus time.

best explained by an example. A first-pass mix of an SBR/polyisoprene blend (as might be used in a traffic counter tubing extrusion) is given in Table 2.22. The procedure is a modified sandwich mix with late oil addition, so as to maximize carbon black dispersion before addition of 'soft' ingredients (which might coat the black and impede dispersion). This order of addition is common when both dispersive work and polymer blending are needed. The batch size is appropriate to a #11 internal mixer, and would be run at 30–35 rpm (with variable speed available) otherwise probably at low speed (in a two-speed machine). The cycle time, allowing 30 s for the batch to dump, would be about 5.5 min and would deliver 4000 lb mixed compound per hour, at an average power draw of 250 kW.

Figure 2.1 plots batch temperature and power draw versus time throughout the batch. The temperature determined using a thermocouple mounted in the drop door will lag behind the actual batch temperature by 5–10 °C; this difference is taken into consideration when listing events on the mix procedure card, events that are to occur based on temperature. Sidewall thermocouples tend to be much less reliable and are best used as backup, alerting operators to failure of the primary thermocouple (when it suddenly lags behind the sidewall secondary). The temperature output on the mixer console is easily split to remote locations (e.g. the chemist's office) with compensation, if needed, for any line loss.

Power draw may be monitored by commercial instruments, or the ammeter reading on the mixer console split to the same location, both outputs driving a two-pen millivolt *X–T* recorder or analog device feeding a microcomputer. (If no ammeter is present, an induction ammeter can be placed around a lead to the drive motor.) Assuming the rotor speed is more or less constant throughout the cycle, the power draw is proportional to the observed current flow. (Changing the rotor speed during the mixing cycle is generally used only as a last resort because it is difficult to achieve consistently from batch to batch, and it increases wear and tear on the power supply.)

The batch appears to be at about 125 °F when loading is complete; this results from room temperature ingredients being loaded into a mixer with its circulating coolant set at about 180–200 °F. (A good starting point, if temperature control is available, is to set the coolant temperature about 100 °F below the expected temperature when the batch is dumped.) A strong power surge (300 kW) is noted as the SBR, carbon black and IR are mixed (from 0.5 to 1.5 min), with a rise in temperature to about 165 °F. The temperature is rising fairly slowly because the partial batch size is small; nonetheless, the three ingredients incorporate in about 0.5 min (the power draw levels off). The power draw drops sharply as the ram is lifted for the second addition, accompanied by a slight temperature drop. The second addition, at 1.5 min into the mix cycle, has now brought the batch close to optimum size. When the ram is lowered, both the power draw and the temperature increase more rapidly than after the first addition.

After another 0.5 min the power draw levels off again and begins to drop. The batch temperature continues to rise. These characteristics typify completion of the incorporation/distribution phase of mixing and the onset of compound viscosity reduction. At 3 min into the cycle, the ram is raised for the final addition, again causing a sharp drop in power draw. Upon addition of oil and waxy ingredients, incorporation and distribution are slow (as the liquids decrease the effective shear rate). When incorporation of the oil and waxes is complete (approaching 4 min into the mix), the power draw levels off at about where it was before the addition, and only a minor increase in batch temperature is noted (from 240 to 250 °F).

At this point the ram was raised and any material on its top surface scraped into the batch. (The traditional jargon is 'swept down', but brooms, or other tools that might lose bristles, should not be used; compressed air can be used if dry and clean, not always the case. Zinc-plated, or otherwise static-free, metal scrapers are usually employed.) When the ram is lowered onto the now complete batch, the power draw (and the compound viscosity) decrease as the temperature builds; the actual batch temperature at the 285 °F dump was about 290–295 °F.

Revision of this mix cycle for improved output depends on the application and the downstream process. Assume, for example, that the compound is indeed to be used for extruded traffic counter tubing. Suppose it is to be accelerated on a two-roll mill and strip delivered to

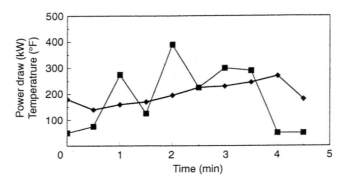

Figure 2.2 SBR/polyisoprene blend, ram down at 0.5 min, second addition at 1.5 min, third addition at 2.5 min, dump at 4 min: (◆) batch temperature and (■) power draw versus time.

a hot feed extruder. A smaller reduction of compound viscosity might be satisfactory, perhaps even an improvement as regards flat spotting or distortion if the compound is extruded unsupported, coiled in pans and autoclaved. On the other hand, if the compound is second-pass mixed and stored as strip before delivery to a cold feed extruder, perhaps the activity of the cure system might be too great to survive the higher levels of shear in the extruder, levels that would accompany increased compound viscosity. In either case it might be possible to obtain grades of the starting elastomers having lower Mooney viscosity, decreasing the need for viscosity reduction during mixing.

A revision that would increase output, but decrease the extent of viscosity reduction, is given in Fig. 2.2. The time between the first and second addition of ingredients has been cut by 15 s. This is the time period between 1.25 and 1.5 min of Fig. 2.1, where the initial incorporation seems complete and the power draw has turned level. A second 15 s has been dropped between the second addition and the last addition of ingredients, the time corresponding to the interval between 2.75 and 3.0 min in Fig. 2.1. Once again, this is a place where the power draw has leveled off. The final 'sweep' has been eliminated, saving another 30 s, assuming that very little unmixed compound was found on the top surface of the ram. This is often the case with multiple ingredient additions to a well-maintained mixer (ram fits properly), particularly when oils and waxes are added with the ram raised (decreasing the tendency to force fluids over the ram).

The total saving is 60 s, leading to a steady production rate of 4800 lb h^{-1}, at an increased average power draw of 275 kW. The increased output is worth 5–10 times the increased power cost (assuming it contributes to earnings, not always a reasonable assumption in the rubber industry).

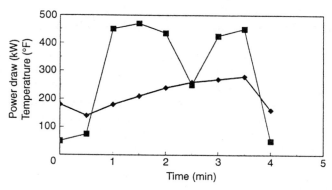

Figure 2.3 SBR/polyisoprene blend, ram down at 0.5 min, sweep at 2.5 min, dump at 3.3 min: (◆) batch temperature and (■) power draw versus time.

The compound resulting from the revised mix procedure, unless based on lower viscosity elastomers, will certainly have higher viscosity than the original. This may or may not be of significance. As only periods involving the lowest levels of power consumption were cut, the physical properties of the compound have probably not been greatly affected.

Suppose it is now decided to use the same recipe to produce compound for compression molding of sink stoppers (rather than adding other raw materials to the inventory). In this case it may be that a lower level of carbon black dispersion is adequate. Then the compound might be mixed by addition of all ingredients at the start of the batch, using a sandwich load. The mix cycle is depicted in Fig. 2.3: sandwich load requiring about 30 s; incorporation and distribution completed in 2 min of intense mixing; a sweep of unmixed compound at 2.5 min into the cycle; and dump at 3.3 min (280 °F). This revision yields about 6000 lb h^{-1}, a very substantial output for this size of mixer, at an average of 320 kW power consumption. The product may be quite useful for molding of miscellaneous tough, elastomeric objects (even if the extreme durability required in traffic counter tubing is not provided).

The usefulness of such a high output revision depends not only on the application, but on whether the mixing equipment can deliver average power input of ≥ 300 kW, with surges of over 450 kW, to a 400 lb batch. The data in question were obtained using a #11 mixer powered by a 1500 hp DC drive motor. When the same batch was mixed in a #11A, driven by a 500 hp AC motor at a rotor speed of 26–28 rpm, a maximum of 180 kW could be obtained. Using the original mix procedure, output was 2000 lb h^{-1} at an average power consumption of 130 kW. Output of mixed compound is not necessarily strictly proportional to power input over a broad range of compound types, but it is a reasonable first approximation to keep in mind when assembling components.

This example was selected not because it typifies the average mixing of rubber compounds, but rather to illustrate how the time–temperature and time–power draw curves may be used to modify mix procedures. The underlying assumption is that those periods when the power and temperature response of the mix seem to be the most undramatic should be the first periods to examine for possible cuts. But this too is merely a first approximation.

REFERENCES

1. Barton, A. F. M. (1983) *Handbook of Solubility Parameters*, CRC Press, Boca Raton FL. Uses SI units.
2. Allied-Signal Specialty Chemicals (1978) Solubility Parameters of Solvents and Polymers, *Report G-7*. Uses hildebrands.
3. Grossman, R. F. (1989) Lubricants, in *Thermoplastic Polymer Additives* (ed. J. T. Lutz), Marcel Dekker, New York.
4. Johnson, P. S. (1985) Mixing equipment and the mixing process, in *Basic Compounding and Processing of Rubber* (ed. H. Long), Rubber Division, American Chemical Society, Akron OH.

3

Mill mixing

Richard F. Grossman

3.1 INTRODUCTION

Mixing rubber compounds on a two-roll mill is common in the following situations:

1. First-pass mix is accelerated, typically by a fabricator, as part of the preparation for molding, calendering, etc.
2. Small batches of specialty compounds are mixed.
3. Compound is accelerated on the mill after being dropped from an internal mixer.
4. Entirely ordinary rubber compounds are mixed on a mill for no other reason than the equipment exists and has always been present.

Other reasons of convenience may also pertain: mills are consistent with low ceiling height restrictions, etc. Most of these make little sense in the larger framework of optimizing output at consistent high quality. The optimum equipment and procedure strongly depend on the particular objective, as well as on the types of compounds to be mixed.

3.2 ACCELERATION OF FIRST-PASS COMPOUND

A common situation involves a custom molder buying mixed compound, intentionally missing all or part of the curatives, then completing the batch on a mill, before extruding strip to feed injection presses, or preforms for compression or transfer molding. The unaccelerated compound usually has a very long shelf life, permitting flexibility in the quantities ordered and safety in the saving of remnants for future use.

The Mixing of Rubber
Edited by Richard F. Grossman
Published in 1997 by Chapman & Hall, London. ISBN 0 412 80490 5.

In addition, the cure rate can be varied (within certain limits) to accommodate seasonal differences and press-to-press variation. Furthermore, blends can be made for minor jobs and workaway can be facilitated. As a result, even fabricators who routinely use fully accelerated compound often keep one or more mills in their plant.

For these purposes, mills having roll sizes of 16 in × 42 in (diameter × length), 18 in × 48 in and 22 in × 60 in are popular. The corresponding batch sizes are often estimated by the following bit of folklore: for roll diameters less than 20 in, the roll width multiplied by the compound specific gravity equals the batch size in pounds; for roll diameters greater than 20 in, use 2 × roll width × specific gravity. Thus with a 16 in × 42 in mill the batch size should be 30–50 lb × specific gravity; with an 18 in × 48 in mill it is 40–75 lb × specific gravity; and with a 22 in × 60 in mill it is 100–140 lb × specific gravity. The bottom of the range should be used with compounds where a thin band provides easy milling, e.g. NBR, EPDM and many specialty elastomers. Conversely, the top of the range would be more common with SBR, natural rubber or butyl. The size of the batch may also be influenced by subsequent operations, e.g. the need for multiple small additions to a calender to avoid overloading and air trapping; or by previous operations, e.g. the weight of a given batch (or blend) of a first-pass mix.

Mills for curative addition should have low friction ratios, typically 1.05–1.2, and should always be installed so that either roll may be worked. Both sides should have controls and multiple safety switches. Regardless of type, the safety switches must provide instantaneous braking, with frequent preventive maintenance checks. Mills in the 42–60 in range are particularly dangerous because they often do not inspire the respect and caution accorded to larger pieces of equipment. The mill rolls are most economically cooled by recirculation of water through refrigeration equipment (generally called a 'chiller'). Even if locally available water is free, it is still advisable to recirculate; it gives better temperature control and it avoids having to dispose of wastewater to the environment. Corrosion inhibitors must be added, and in-line filters routinely serviced.

Three types of water-cooled mill rolls are usually available: rolls having many drilled passages for high thermal transfer; rolls that are hollow and in which the internal cavity is sprayed with cool water; and rolls that are hollow but in which cool water merely sloshes around. The first type, 'drilled' rolls, are excellent but expensive; they are generally used in the cure addition application for sensitive compounds only. (They are much more widely used in sheet-off mills for one-pass batch mixes.) The second type, analogous to the 'spray-side' internal mixer, is adequate for most cure additions. But frequent preventive maintenance is required to ensure the spray nozzles are operational. When the nozzles become plugged the cooling degenerates and becomes similar to the third type (water sloshing inside a hollow roll); this is useful only with the most forgiving compounds and it should be avoided if possible.

For this type of application, the drive motor should have about twice the horsepower as the width of the mill in inches. A 42 in mill is typically driven by a 75 hp motor; 48 in mills by 100 hp motors; and 60 in mills by 125–150 hp motors. If a variety of compounds are processed (as is increasingly the case), it is very helpful to install a variable-speed DC drive. In the above horsepower range, silicon controlled rectifier (SCR) drives are convenient (although motor–generator drives remain more common). With variable speed it is not only possible to maximize coordination with downstream handling and processing, but also to vary the overall shear and heating delivered to the mill batch (section, 1.1). As a safety measure, gear reducers and the overall drivetrain should naturally be sized with greater capacity than the horsepower of the drive motor.

The most ingratiating installations are 'unitized'; the various drive components are fixed to a frame, facilitating transport and relocation. That is not to say that such movement should be carried out without realignment of the components, instrumentally, by qualified technicians. Failure to attend to these details will decrease the life of the connecting and drive gears, the most expensive components to replace. A further common precaution is to insist upon flexible couplings between components, even in unitized constructions. Preventive maintenance should include periodic checks on alignment, and they should be carried out by professionals. Once the rubber technologist begins to hear strange sounds or to notice undue vibration, some parts may already have suffered considerable damage.

With 42 and 48 in mills the operator most often hand-cuts pieces to feed a preforming extruder; 60 in or larger mills often use strip-cutting knives. These operations are generally slow enough to work satisfactorily with knives that swing in from one side. Alternatively, a bar may be dropped into place (on either side of the mill) bearing spring-loaded knives for strip cutting. As with mill knife blades, these should be of ordinary carbon steel to permit frequent sharpening and to minimize scoring of the mill roll's hardened surface. Regardless of the size of the equipment, strip cutting should never be assisted by running the mill with the rolls other than completely parallel. The alternative, 'cocking' the rolls, will drive the bank of rubber in one direction or another (depending on whether it prefers steel adhesion or self-adhesion); it will also prematurely destroy the connecting gears.

The cure addition operation seldom requires motorized hydraulic roll nip closure; manual operation is usually adequate. All two-roll mills used in a production environment, regardless of size, should be equipped with automatic lubricating equipment and an alarm for indication of insufficient lubrication. (This should be less strident than the sound that accompanies a mill safety switch; the safety alarm should be so piercing that no operator would contemplate routine shutoff by use of the emergency safety instead of using the 'off' button.)

In actual operation, first-pass (unaccelerated) compound is pre-weighed, often with crossblending of batches based on previous test results, and banded on the mill. If more than one or two batches are to be accelerated, the best practice is to make a blend of the curatives on a scale large enough to accommodate the entire run, e.g. with a ribbon blender, and to accelerate a laboratory sample to confirm test properties. Alternatively, such pretested blends may be purchased from dispersion suppliers, eliminating the need to stock diverse and often irritating chemicals. Mill mixing temperatures may sometimes be high enough to add curatives in low melting bags, bag and all. This is sometimes a possibility with autoclave or steam continuous vulcanization (CV) cures. With most molded products, the bag must be emptied into the mill nip then disposed of. If such packaging material cannot be returned to the supplier, it may often be fluxed into cleanout batches and thereby converted to nonhazardous solid waste, at least in most cases.

Table 3.1 NR/SBR belt cover

	First pass: #11 internal mixer
Polyisoprene	125 lb
SBR 1502	75
BR 8407	75
N347	125
Zinc oxide	10
Zinc laurate	1
Antioxidant	4
Aromatic oil	10
Total	425
	48 in mill batch[a]
Batch	85 lb
Sulfur	1.2
OBTS	0.4
DPG	0.4

[a]Last batch may be smaller.

Table 3.2 NBR injection molding compound

	First Pass: #11 mixer
NBR, medium ACN, medium Mooney	150 lb
NBR, low ACN, medium Mooney	100
N550 FEF	100
Hard clay	50
Phthalate ester plasticizer	50
Factice	25
Zinc oxide	7.5
Antioxidant	5
Treated sulfur	2.5
Total	490
48 in mill batch[a]	
Batch	70 lb
MBTS	0.7
TMTD	0.4

[a]Last batch may need correction.

The goal is to convert the curative addition step into an operation easily tracked by statistical process control (SPC). A useful adjunct is to station test equipment, such as a rheometer, next to the cure addition mill, so the mill operator can perform quality control analysis. This reinforces the importance of accurate weighing and consistent working of the compound on the mill. Most curatives disperse easily in common compounds. All that is usually required is to add the cure ingredients to the rubber traveling on the roll into the nip, sweep the powders that fall through once or twice from the mill pan, often into a common dustpan, for readdition before cutting the batch from side to side for thorough blending.

Sulfur is often an exception. In many compounds, sulfur is therefore added to the first pass. Where high dump temperatures, or the need to incorporate some other intractable curative in the first pass, militate against doing this, it is often advisable to add sulfur on the mill in a predispersed form, e.g. as a paste in process oil or a dispersion in the same type of elastomer. Custom dispersions can often be made that introduce no new ingredients into the recipe. For example, in the compound of Table 3.1 mill addition of the cures would best use sulfur in dispersed form. On the other hand, in the NBR compound of Table 3.2 most technologists would add sulfur in the first pass and use mill addition at the fabricator's site to mix in the accelerators.

3.3 MILL MIXING OF SPECIALTY COMPOUNDS

What constitutes a specialty compound depends upon one's point of view. Compounding experimental elastomers, fresh from the polymer synthesis laboratory, is often assisted by small tabletop mills, commonly having roll sizes of 3 in × 6 in, 3 in × 7 in or 3 in × 8 in and driven by 1.5–2.5 hp motors (small enough to run at variable speed by several methods). These mills are useful for 100–250 g batches, but with care they can also be used to add ingredients to 50–100 g batches of experimental polymer. In addition, the 3 in × 8 in size is suitable for sheeting off the 350–400 g output of the smallest common (Midget) laboratory internal mixer. These miniature mills are also very useful for checking proper dispersion of ingredients in samples of compound mixed with larger equipment.

When used with experimental polymers of undetermined stability and unknown sensitivity to trace metals, it is best to equip the miniature mill with chrome-plated rolls. This is contrary to general-purpose use, where chrome plating provides a visually attractive surface that is very brittle and easily damaged. With chrome-plated rolls it is necessary to insist upon relatively soft mill knife blades, which are unpopular because they rapidly lose their knife-edge, and on frequent inspection of the stops that prevent inadvertent roll-to-roll contact if the nip setting is too tight.

The most common small mill is the 6 in × 12 in or 6 in × 13 in model found in almost every laboratory. These mills, usually driven by 7.5 hp motors, are most often two-speed, less often variable-speed. Commonly used to sheet off batches from a lab internal mixer of 2–3 lb capacity, they are also well adapted to mixing experimental compounds on a 1–2 lb scale. With a tight nip setting, even normally mill-intractable elastomers, such as semicrystalline EPDM, can be mixed in small quantities. But it may still be much more preferable to mix these compounds in a laboratory mixer; however, laboratory mixers are several times more expensive to purchase. Small quantities of special-purpose elastomers, such as phosphonitrilics and fully fluorinated polymers, used to mold small parts, are often mixed commercially with lab mills.

Somewhat larger sizes, 8 in × 16 in (usually 15 hp) and 10 in × 20 in (usually 20 hp) pilot-scale mills are often used to sheet off batches from the common 7–9 lb pilot plant internal mixer. Complete batches can also be mixed, 3–4 lb on the 16 in mill and 5–8 lb on the 20 in mill. A common use for the next sizes up, 12 in × 24 in (usually 40 hp) and 14 in × 30 in (usually 50–60 hp) is in sheeting off the output of the 27–32 lb pilot-scale internal mixer. This combination provides a batch large enough to mold a test tire tread and is therefore popular. The 24 in and 30 in sizes are also well suited for small batches of expensive silicone or fluoro-elastomer compounds.

Starting-point equipment conditions for mill mixing of specialty elastomer compounds may be estimated by considering related procedures for internal mixers; many of them are covered in Chapter 8. A reasonable first approximation is that roll temperatures should be similar to the settings of the sheet-off mill used with an internal mixer. Underlying this is the premise that the batch temperature during the last minute or two of internal mixing represents a good choice for optimum distribution and dispersion of aggregates, as well as for completing the incorporation of ingredients that are slow to wet. This is usually the case. There are, however, several specialty elastomers where incorporation and distribution are best accomplished early in the cycle. These are generally elastomers with molecular weights much lower than, for example, natural rubber or SBR. The mixing mill rolls must then be set at temperatures lower than would normally be used on a sheet-off mill for the same compound.

Mill mixing involves long ambient exposure at elevated temperatures, so condensation polymers need to be protected against hydrolysis. Rubber storage should be at temperatures well above the ambient dew point. Early addition of trace quantities of a water scavenger, such as polycarbodiimide, should also be considered. Furthermore, most recipes should be adjusted to a higher level of antioxidant content than would be used with internal mixing. Curatives are best added as dispersions or cure packages preweighed to the batch size. This minimizes variability in mill pan loss.

In many operations all small powder ingredients are preblended and weighed into packages for each batch. Alternatively, such packages are available commercially. There is almost nothing available in the literature regarding shelf stability of such combinations; the best course is prompt consumption. If interaction is suspect, suppliers of cure packages will segregate ingredients into multiple containers.

The least attractive feature of mill mixing is the addition of liquid ingredients. Unless carried out extremely slowly, the effect is to break the adhesion of the batch to the front roll, interfering with incorporation of ingredients. Simultaneous oil and filler addition is common in the laboratory, whereas for production it is usually convenient to preblend fillers and liquids before mixing. It is useful to attempt a formulation so the result is more or less a wetted powder. If such blending cannot be done in-house, this type of service is available commercially. Another alternative is to purchase one or more masterbatches from custom compounders and to combine them on a mill, perhaps with addition of curatives. In this way the overall secrecy of the formulation can be maintained while others carry out the more cumbersome mixing steps.

3.4 ACCELERATION IN LINE WITH INTERNAL MIXING

At a number of sites, internal mixing is combined with addition of curatives on a mill in the same production line. There are several schemes for this process, depending on the temperature at which the batch is dropped from the internal mixer. If this exceeds a safe temperature for cure addition, the batch must be cooled. A common guideline is not to come within 100 °F of the standard vulcanization temperature. This will vary with the type and level of curatives, and the sensitivity of the elastomer. Cooling is carried out by dropping the batch and working it on a cool mill. This mill can be equipped with overhead blender rolls for additional cooling; they simply extend the batch surface in contact with chilled rolls. In this operation it is desirable to use a mill having drilled rolls (section 3.2) for best cooling. If the compound has good scorch safety, cored rolls with water spray may be sufficient. It is essential to spot-check batch temperature during the cooling process; a permanently mounted infrared thermometer is an excellent investment here.

The cooled batch is usually transferred to a second mill for cure addition. This enables cure addition to one batch while the succeeding batch is cooled and a third batch is mixed. And the first mill is kept free of loose powder that can interfere with efficient handling of batches dropped from the internal mixer. The second mill will usually have gear ratios providing a higher friction ratio than the cooling mill; this is to facilitate rapid incorporation and distribution of the curatives. The second mill may also run at higher speed, and it may also use an overhead blender if continued cooling is needed.

After leaving the second mill, the compound is sometimes transferred to a third mill for forming into the desired dimensions for packaging. This system derives from traditional operations that used a series of three (or more) mills instead of internal mixing. Many such installations were upgraded by addition of an internal mixer.

4

Quality control and the mixing process

Richard F. Grossman

4.1 INTRODUCTION

Some of the factors that repeatedly enter into the mixing of rubber compounds are relatively straightforward: the raw materials used must be consistent; the same mix procedure must be followed, with the same heat and shear history throughout the process; contamination must be avoided. Other factors are less obvious, but nevertheless important: the compound should be formulated so as to minimize the effect of small variations in processing, and to minimize the chances for errors in composition. The type of mixing procedure (the order in which the ingredients are added) should be appropriate to the formulation of the ingredients and to the design of the internal mixer, so as to minimize the dependency of compound properties on small process variations. Finally, the mixer itself must be well maintained (clearances within specifications, proper circulation of coolant, etc.) for a consistent product to be obtained. This chapter will be concerned with measurement of the consistency of raw materials and finished compound.

4.2 TESTING OF RAW MATERIALS

Consistent mixed compound cannot be obtained without testing each lot of every raw material. Such a program is invariably expensive. The consequences of not testing inevitably prove to be more expensive. Every raw material should be examined visually. This will eliminate most cases

The Mixing of Rubber
Edited by Richard F. Grossman
Published in 1997 by Chapman & Hall, London. ISBN 0 412 80490 5.

of contamination (except for inadvertent contamination caused later on). Merely looking at the ingredient before using it will identify some defects, such as polymers with oxidized surface layers. Many cases of mistaken identity (or mislabeling) will be similarly spotted. The raw material specification for every ingredient should include visual inspection.

Each lot of incoming raw material should include certified test data for that lot, signed by a responsible individual on behalf of the supplier. The certification should contain actual test data, not merely a statement that the shipment complies with the supplier's sales specifications (the latter often furnished with 'typical' test data). What is desired is specific data that can be used for statistical tracking, and repeated on checking. With many ingredients in routine use, the combination of visual examination and perusal of the batch certification will prove sufficient. Comparison of individual certifications with the published specification on the ingredient should not be ignored; many cases will be found where lots deviating from the supplier's specification have been inadvertently (or otherwise) shipped. Judgment will be needed in such cases to decide whether a shipment must be returned, or may be used (perhaps in one compound, but not in another).

4.2.1 Elastomers as raw materials

Incoming polymers should have their certification carefully checked for values of ash and volatiles. Unusually high values should be questioned. High ash should not reflect incursions of dirt or grit. High levels of volatiles may correlate with a moisture content great enough to make filler incorporation difficult. If in doubt, sample the polymer and run a test batch in the laboratory under close observation.

In cases where compound processability is critical, visual inspection and audit of certified data should be supplemented by a check of polymer viscosity. It is usually the Mooney viscosity that is compared with the supplier's data. For this to be meaningful, it must be established that exactly the same test procedure is used. ASTM D1646 (Mooney viscosity of natural and synthetic rubbers) includes several variations. It is necessary to use exactly the same sample preparation to duplicate results. And the test instruments should be calibrated by analogous methods. Some discrepancies have actually been traced to use of the wrong sized rotor. More often the lack of correlation stems from a worn rotor or worn dies, inadequate temperature control, and variation in sample preparation. A former source of variation, die spacing, has been minimized by modern design, but will still affect older, hand-closed Mooney viscometers. Other viscometers or plastometers can also be used; in fact, extrusion plastometers are more appropriate for compounds that are later to be extruded. In these cases, only internal correlations will generally be available (unless a polymer supplier can be persuaded to provide corresponding data). The largest sources of rejection of incoming elastomers are contamination and improper viscosity.

4.2.2 Fillers

Inert fillers (those that do not reinforce) should be inspected visually for contaminants. Inspection of incoming polymers is generally simple: corners of bales are cut off here and there (following a predetermined statistical frequency). Even if the polymer is powdered or pelletized, a bag can be partially opened and resealed without disrupting normal housekeeping. If fillers are used in bulk, samples can be obtained during transfer to silos or holding bins. On the other hand, if fillers are used in bags, judgment is needed to balance the need for materials inspection versus plant cleanliness. Where the ingredient has a record for consistently being as stated, inspection is often deferred until bags are opened just before mixing. It may sometimes be useful to assign such quality inspection roles to scalemen and materials handlers, thereby increasing their involvement with product quality. But if there is reason for suspicion, then bags must be opened (and resealed) for sampling, regardless of inconvenience.

As with polymers, reported ash and volatiles should be scrutinized. With reinforcing fillers, some measure of particle size, shape and distribution should be noted. With reinforcing carbon blacks, iodine and dibutyl phthalate (DBP) numbers can be tracked. With nonblack reinforcing fillers, actual particle size distribution for the lot should be reported by the supplier. Variations in these parameters should be investigated to discover their effects on the compound's set of physical properties. Significant changes in reinforcement behavior generally modify the entire set of properties: modulus, hysteresis, resilience, etc., rather than a single property.

Fillers not of neutral pH often affect the cure rate and sometimes the ultimate state of the cure. Cure systems that are pH sensitive should then be avoided, but this may not always be possible. With fine particle precipitated mineral fillers (silicas and silicates), it is often useful to check the pH of an aqueous slurry. Most of these products look alike, but the substitution of an acidic grade for a neutral analog could prove disastrous in some cases.

By far the largest cause of rejection of individual lots of filler is contamination with foreign objects. This varies greatly with the type of filler, being relatively more common (though, that is, still uncommon) with mined fillers, less so with more expensive precipitated or fumed types.

High moisture content is a less common defect, but it proves serious when encountered. With batches that mix with high shear rates and are mixed to high dump temperatures, rational mix procedures can often accommodate levels of filler moisture that are higher than normal. On the other hand, it simply may not be possible to disperse a moist filler into a soft elastomer at modest temperatures. Raw material acceptance must therefore correlate with established compound design and mixing capability.

4.2.3 Plasticizers and process oils

Plasticizers and process oils tend to exhibit subtle differences in appearance and odor, recognizable to the experienced. If such methods are relied upon exclusively, it is certain that the wrong ingredient will eventually be used. Upon arrival, each fluid should be checked before acceptance (particularly if it is to be pumped into a tank or reservoir). A plasticizer or process oil that has the correct refractive index (ASTM D1045) and an infrared spectrum identical to a standard (ASTM D3677) is almost surely the anticipated ingredient. Both tests can be run in the time needed for the arriving teamster to consume a cup of coffee. If testing is omitted, it is certain that one day the process oil tank will inadvertently be filled with bunker oil or kerosene.

4.2.4 Small ingredients

Organic ingredients (accelerators, antioxidants, activators, lubricants) that are solids are quickly identified by melting point (ASTM D1519). If any uncertainty arises, or in the case of liquid ingredients, the infrared spectrum may be compared with the spectrum of a standard sample. Mislabeling of such ingredients is rare, but it may have catastrophic effects on the finished compound. Use of the wrong curative may be subsequently diagnosed in testing the compound, but use of an inappropriate antidegradant could easily escape undetected (until field failures ensue).

Additives that are largely inorganic, such as lead-based stabilizers, can be checked conveniently by ash determination compared with a standard (ash recovery is rarely 100% of theoretical) per ASTM D297. Solids that are completely inorganic (zinc oxide, antimony oxide) are best considered analogous to fillers when it comes to raw material acceptance, except where particle size is critical. Records of past performance should indicate whether to accept the certified data or to perform a test (e.g. via sieve analysis as per ASTM D1210).

Despite the best intentions in compound design, instances will arise where small variations in a key ingredient disproportionately influence compound properties. Examples include activity of organic peroxide cure initiators and of cross-linking monomers in some compounds, usually those stretched to the furthest limits of extension and dilution. Sometimes it may be necessary to run laboratory batches with samples and even to revise the level that is used. Under these circumstances it is often possible to persuade the supplier to provide preshipment samples for lot approval (but these must be rechecked at a later date, at least occasionally).

In the long run, any apparent statistical correlation of variation in compound properties with raw materials is of even greater importance than immediate problems of mistaken identity, for it makes possible rational, rather than intuitive, revision of compound design.

4.3 CONTROL OF COMPOSITION

Consistency of composition should be considered as far back as the laboratory, where the recipe originates. Two factors are involved: the units of weight in which the ingredients arrive, and the optimum batch size of the mixing equipment that will be used. Underlying these factors is the premise that the fewer the metering operations to be carried out, the fewer the random errors in composition. This is true regardless of the methods of metering. Automated methods offer the potential of great precision, but they introduce their own, often sophisticated, sources of error. This is not to argue against automated methods – with proper calibration and maintenance, very good results may be obtained – but simply to note that automation of weighing and handling is not, in itself, any guarantee of compositional accuracy.

In the ideal case, the formulation consists of a limited number of ingredients, all delivered from bulk storage, where numerous lots of incoming raw materials have been crossblended (after acceptance testing) to minimize variation. Such a situation is very nearly the case in some high technology fields (e.g. components of radial tires). Most commonly in the rubber industry, however, elastomers are used in the bale form in which they are supplied; fillers in bags; small ingredients weighed from the supplier's package.

The technologist should begin recipe development thinking of the batch size of the mixer. That is, he or she should begin not with 100 parts of polymer, but with the quantities of elastomers that correspond to the optimum size of an anticipated batch. At the start, this involves a certain amount of intuitive thinking. For example, it may be desired to develop a relatively low cost 60–65A durometer SBR compound based on SBR black masterbatch. The latter is routinely supplied in 90 lb bales (originally intended for the mixing of one bale on a 60 in two-roll mill). For a #11 or F-270 Banbury mixer, the compounder should then begin

Table 4.1 Formulation revised for efficient weighing

	Original	*Revised*
EPDM (75 lb bales)	70.0	75.0
EPDM (55 lb bales)	57.5	55.0
Butyl (75 lb bales)	27.5	25.0
Whiting (50 lb bags)	210.0	200.0
Calcined clay (50 lb bags)	105.0	100.0
Hard clay (50 lb bags)	87.5	100.0
Process oil	40	40
Small ingredients	24	24

with 270 parts of SBR black masterbatch (or 180 parts for a #9, 360 parts for an F-370, etc.). This assumes that the consistency of bale weight (as supplied) is adequate for the application. In most cases the typical bale weight variation (usually a maximum of ± 1 lb from theoretical) is acceptable; in a few cases it is not.

Table 4.1 compares actual #11 batches for an EPDM/butyl blend, originally developed from parts per hundred of polymer, with a revision in which ingredient weights are rationalized. Reduction in the frequency of cutting bales correlates strongly with compositional accuracy; that is, the elastomer supplier is usually more accurate than the processor. The one instance of bale cutting needed in the revised compound, dividing a 75 lb bale of butyl rubber in thirds, can generally be managed within a 1 lb tolerance. Cutting bales in halves or thirds is invariably done more accurately than more complex divisions.

A further advantage is that the revised compound can be delivered to the mixer as a row of future batches, all of the same appearance, starting (or ending, depending on the mix procedure) with two (different) EPDM bales and the same sized cut piece of butyl. In this way the mixer operator may be incorporated into the quality control staff, being trained to alert a supervisor if the ingredients for a given batch are different in appearance or configuration from preceding batches. This function is not possible if batches to be mixed are based on a motley array of cut pieces of rubber of different size and shape. And a motley array of sizes commonly means that small pieces fall from the batch in transit to the mixer, often to be misplaced in an adjacent batch.

The revised compound of Table 4.1 also rationalizes filler usage into an even number of bags (unnecessary when fillers are delivered from bulk). As with bales of rubber, such rationalization is not always possible, even when beginning with a new recipe. Nonetheless, in the example given above, the compounder might well consider 270 lb SBR black masterbatch plus 100 lb mineral filler as a starting point for a #11 mixer, 360 lb plus 150 lb for an F-370 etc. The same numbers, in grams, can usually be multiplied by convenient factors for experimentation in a B or C laboratory Banbury or two-roll mill.

A common arrangement for delivery of unmixed batches to the mixer comprises a conveyor bearing an assembly of completely distinct batches, each of identical configuration. Each batch may be accompanied by a weigh ticket, containing confirmations from automated ingredient delivery devices, or initials by scalemen, reflecting addition of proper weights of components. This assembly can be inspected rapidly and efficiently by quality control technicians, often with random sampling to confirm ingredient identities. Automated delivery equipment may also update remote locations of weighing actions, alerting supervisors to potential problems.

In some installations the revised compound given in Table 4.1 would be mixed by delivering full units (e.g. 2000 lb containers) of the two EPDM polymers, 25 lb cut pieces of butyl rubber, full units of fillers and pans of preweighed small ingredients to the immediate vicinity of the mixer. The batch is then assembled just ahead of mixing. (The process oil would normally be injected from a weigh tank.) This procedure tends to conserve space and promotes high output (many batches may be mixed more rapidly than they are assembled), but it also increases the incidence of compositional errors by limiting the opportunity for inspection of the unmixed batch. Mixing facilities concerned with compounds for critical applications have generally abandoned procedures of this type in favor of extensive automation of weighing and batch assembly, and increased frequency of inspection before mixing. This is a prerequisite for statistical tracking of errors in composition, a vital factor if analysis of the effects on the compound of raw material variation is to have any meaning.

Control of composition is another factor during the mix, insofar as direct metering of ingredients to the mixer is used. As noted earlier, this is most common with process oils. Direct injection of oil improves output, provides closer control of oil incorporation, and is a vast improvement in housekeeping over addition via buckets or pails, as well as increasing the accuracy of weighing. Like automated features in general, this is true only as long as the injection device is suitably calibrated and in working order. Furthermore, the apparatus must be matched to the viscosity and solvent power (effect on gaskets and seals) of the fluid or fluids to be injected; injection settings appropriate for heavy oils will not accurately dispense low viscosity plasticizers. To accommodate practical situations where several fluids are called for, a manifold is often used to meter oils and plasticizers from several sources into one or more weigh tanks, from which direct injection is accomplished using adjustable devices. In concert with this, high viscosity fluids are typically heated to temperatures at which they are readily pumped; this also tends to facilitate incorporation into the compound.

If a batch were to use a large quantity of heavy process oil and a small amount of a low viscosity plasticizer, the oil might be injected, and the plasticizer added using a bucket, simply out of convenience. Hand addition of liquids is never precise and often adheres loose filler to hopper and throat surfaces, increasing subsequent cleaning requirements. These effects are magnified in cases where, through sloth or through misguidedly thinking that output may be increased, the operator fails to lift the ram before adding the fluid. With critical compounds it is more satisfactory to use small amounts of liquid ingredients (that are not readily injected directly) in the form of predispersions on inert solid carriers. This is also the case with extremely viscous fluids that are not easily pumped, such as resins or low molecular weight liquid elastomers. And with such materials, dispersions in powder or paste form often increase the rate of incorporation into the compound.

In a general sense, preblending should be considered as an aid to compositional accuracy, particularly with key ingredients (those whose effects are substantial, even though present in small quantities). An ingredient of a type where small weighing errors can produce disproportionate changes in compound properties may be added efficiently if predispersed in another ingredient (already in the formulation) so as to increase the absolute magnitude of the weight used. Thus it may be advantageous, for example, to add 1.0 lb of a blend of 40% silane coupling agent and 60% paraffin wax, than to install a device to meter 0.4 lb of the pure ingredient. A great number of such combinations can be preblended (or purchased in a preblended state) without introduction of extraneous or unknown ingredients.

4.4 TRACKING THE MIX CYCLE

Unfortunately, many people involved in the mixing of rubber compound believe that, if the correct weights of the proper ingredients are added, and the batch mixed until visually homogeneous but not scorched, the specific details of procedure in going from the unmixed to the final state are inconsequential – at least no more significant than whether cream is added to coffee or vice versa. This attitude usually stems from a lack of appreciation of the polymeric state. It is usually a very poor approximation. During the manufacture of a rubber article, the processability and

Table 4.2 SBR/NBR/PVC blend

NBR/PVC (2 bales)	100 lb
SBR 1848 (2 + 1/3 bale)	210
Hard clay (1.5 bags)	75
Zinc oxide	8
Antioxidant	2.5
Stearic acid	1.5
Aromatic process oil	40
Sulfur, treated	3.5
MBTS	3.5
TMTD	1.5

0 min		Load NBR/PVC
		Load clay, ZnO, sulfur, antioxidant, stearic acid
		Load oil, load cut piece and full bale of SBR 1848
1.5 min		Load MBTS, TMTD and last bale of SBR 1848
~ 4 min	105–110 °C	Dump

the properties of the product have a strong but complex dependence on the heat and shear history of the composition. Variations in this history may not always be detected in laboratory testing of the mixed compound. It is therefore essential to have consistent mixing if analysis of these variations is to be related to ingredients (or any other factor).

The mix cycle is most conveniently tracked by charting both temperature and power draw versus time, as described in section 2.5. The procedure for a common product, calendered sheeting having some degree of oil resistance, is given in Table 4.2. As this compound uses two elastomers of different polarity, SBR and NBR/PVC alloy, a sandwich mix (section 2.4.3) is suggested. The batch was written for a #11 mixer. The batch is loaded into a relatively cool mixing chamber at low rotor speed as relatively active cures will be used in a one-pass mix. After the first loading of ingredients, NBR/PVC at the bottom of the 'sandwich,' and part of the SBR black masterbatch on top, the batch remains almost 100 lb short of optimum size. Nonetheless, the addition of whole bales of rubber causes a fairly strong power surge that carries out incorporation and distribution of the initial ingredients. This surge continues only while the bales are breaking up and massing together with the other ingredients to form the compound. As soon as this part of the cycle is complete, the power draw will level off and start to decline, and the rate of temperature increase will diminish. Close following of the first few batches (when the compound was introduced) led the technologist to set the final addition of ingredients 1.5 min into the cycle. This point (as the final mix procedure is developed) should be characterized by acceptable ranges of batch temperature, current power draw and cumulative power input. For the purposes of the mixer operator, it may be sufficient to specify meter readings. A supervisor can then be alerted if a specific batch displays uncharacteristic power draw or temperature at 1.5 min. For example, if the weight of oil injected was incorrect (from a technical malfunction), an unusual power and temperature profile would almost certainly be noted. At this early stage, a revised procedure could be put in motion (e.g. curatives omitted and the batch set aside).

For statistical analysis, however, it is important to have the continuous monitoring of temperature and power consumption permanently recorded. The signals could both drive a two-pen $X–T$ recorder and input data to a microcomputer. This would be particularly useful if test properties of the finished compound arrived as input to the same machine for analysis of variance.

The remainder of the cycle is tracked in the same manner. Suggested data for each batch might include time and mixer temperature at which initial loading began; maximum power draw in the first stage; time, temperature and cumulative power consumption at the point of the second addition of materials; maximum power draw during the second stage of mixing; dump time, temperature and cumulative power consumption. However, most cases require a smaller number of parameters to control the operation (parameters that prove statistically relevant). The significant variables can then be used to 'close the loop', either literally, with automated cycles, or if this is not the case, by updating the mix card with the observations to be made.

4.5 COMPOUND TESTING

Several tests are typically run on the mixed compound to make sure it will yield the desired properties and processability; and these tests must be run routinely. Their frequency should be determined statistically by considering the precision of the properties needed in a particular application. This may involve non-technical considerations; for example, it may be necessary to guarantee a minimum of 12 MPa tensile strength for a rubber sink stopper, even though it has no bearing on performance.

Examination of the mixed compound is concerned with two other goals: investigating the consistency of the ingredients and investigating the mixing procedure. Sometimes this overlaps with demonstrating the compound's suitability for service, but investigating the ingredients and the mixing procedure is not identical with demonstrating a suitable compound. For example, in most cases the frequency of testing needed for analysis of variance is every batch (because so many potential sources of variation are usually involved). The frequency of testing needed to ensure suitability for service is commonly much lower. There used to be considerable confusion between these two goals, and batches were tested with only 10–20% frequency. In the 1970s it was generally recognized that, to assure consistency, most compounds require 100% frequency of testing. Exceptions are limited mainly to simple master-batches with few ingredients (no curatives) for less than critical applications.

In addition to an unrealistically low testing frequency, many facilities formerly delayed compound testing as convenience dictated. Advancing costs have made such an approach increasingly untenable. The modern mixing facility now performs diagnostic tests on one batch while the next batch is being mixed. This has been made possible by advances in laboratory instrumentation. And for batches containing curatives, it is usually possible to obtain a rheometer curve in the time needed to mix the next batch. This puts a theoretical limit of 2 on the number of consecutive batches that are defective. Sometimes this is clear-cut: a batch yields a rheometer curve indicating no cure; one or more curatives are (most likely) missing. A call to the mixer operator or super-visor can then initiate a hold after (or during) the subsequent batch for a check on ingredients. The defective batch can be isolated (and rechecked) before being comingled with others. The savings are well worth the effort.

The production department charged with efficient mixing is natur-ally reluctant to suspend operations except when clearly necessary. Nevertheless, the cost of rejected compound is far higher than the cost of machine time. During a period of downtime, worth perhaps $500, it would be easy to produce $5000 worth of unusable compound. Down-time used to prevent rejected compound is cost-effective. To be accept-able to all parties, batch rejection needs to be based on indisputable measurements, so the instruments used to test each batch should be regularly calibrated, and properly maintained. Batches cited as defective should be rechecked promptly.

Testing for consistency is greatly helped by maintaining the batch identity; this is often possible, but not always. Difficulties arise when operations downstream of the mixer run continuously. If the individual batches are to be blended before the next step (so as to average out minor variations), they should still be tested before blending. The test data can then be used to assist in blending (which, itself, should be regarded as a unit operation and tracked in its influence on consistent properties).

Matters of convenience, on the other hand, should not intrude on maintenance of batch integrity. For example, motorized hydraulic opening and tightening of the mill nip can often substitute for the practice of leaving part of the last batch on the mill to assist banding the next. Or, if a sizable quantity of a previous batch remains after removal of narrow strip (e.g. for cold feed extrusion), this can be cut into the mill pan then fed into the subsequent batch after a sample has been cut for testing. Whenever a defective batch is found, as a general measure, it is prudent to recheck the previous batch and the subsequent batch. Despite the best intentions, batch identity is relatively easy to lose in a typical mixing room.

The question of whether adequate incorporation and distribution have occurred is often first approached by one of several procedures called **dispersion checks**. A small sample of the batch is commonly passed once through a lab mill at its tightest setting (taking care to avoid actual roll-to-roll contact). The paper-thin compound may be further stretched by hand and examined visually, either with the unaided eye or under 10–20 power magnification (ASTM D2663). One hopes not to see pockets of unmixed powder, specks of clear polymer, contaminants or lumps of one or more ingredients that have agglomerated instead of being incorporated. Alternatively, a rod or tape may be extruded and similarly examined. With critical applications (roll covers, printing blankets, wire insulation) higher magnifications are sometimes used. Both extrusion and mill checks have the advantage over merely examining the surfaces of cuts (made with a scalpel or razor blade) of displaying much more area. In a few cases, even a single pass through a tight mill will tend to incorporate a defect, and examination of a cut surface is preferred.

The most widespread practice is to hold a batch rejected by such a dispersion check for a day to insure complete cooling, then to remix. This is often satisfactory in isolated instances, assuming that further testing does not find other defects. Once remixed and retested, if the batch is now acceptable, it should usually be blended into a subsequent run. Admittedly there will be instances where the parameters of the application, particularly the processing parameters, are broad enough to permit direct use of a batch that has been mixed once more than is standard. There are also cases where no further use can be tolerated. Many manufacturers cultivate certain low-end products merely for disposal of compound that would be marginal in more critical applications.

A batch containing curatives will typically be subject to some kind of dispersion check and will have its processing and cure behavior examined using a rheometer. The results of the dispersion check are an attribute of the batch, but the rheometer curve will yield several variables (ASTM D2084). Those commonly tracked include minimum viscosity, scorch point, cure rate and maximum viscosity. Combining the graph of temperature and power draw versus time with the rheometer curve will pinpoint many compositional defects (excess oil, excess filler, etc.). Experience with the cure behavior of the compound will facilitate interpretation of rheometer curves, reflecting anomalies in the cure system (X was omitted, Y was used in place of X, etc.) The combination of rheometer curves on each batch and analysis of the full property set of the vulcanizates – perhaps run every 10 batches as a demonstration of suitability for service – will provide the background needed to diagnose many other deviations.

With compounds not containing curatives, Mooney viscosity and a dispersion check are commonly run on each batch (other measurements of processability may be used instead). Depending on the criticality of the application and the history of variation, the cures are then added to every fifth, or tenth (or other interval) batch, and a set of properties determined. As with the other quality control measures, they should be considered typical of the rubber industry in general, but not necessarily applicable to a particular case.

Compound testing should not be considered complete when the batch sample yields acceptable test results. Testing should include checks of whether the batch was packaged in the correct form (strip or slabs of the proper dimensions, properly formed pellets, etc.), adequately cooled (and dried, if water cooling was used), and correctly release-coated (if release agents are required). Use of the right weights of the intended ingredients, and close following of the established mixing procedure are meaningless unless the batch is processed with the same care after mixing.

Postmixing checks should also include verification that the batch is correctly identified, another absolute prerequisite for meaningful analysis of data. The quality control report should include specific items to be checked off (e.g. initialed) as part of the test program (temperature of the batch when packaged, correct identification, etc.). These items should include a visual check for any contamination. Inspection of the raw materials will not detect contamination that occurs after mixing. It is usually not feasible (except for tire components and similar products run in vast volume) to dedicate a process line to the mixing of a single compound. Other compounds of similar appearance are usually present. Well-managed facilities provide separate storage areas for recently finished compound (pending quality acceptance), first-pass compound requiring further mixing, and completed compound in sealed packages.

It is often useful to rotate operators and supervisors from the mixing department to areas involved in downstream processing, so that mixed compound can be identified with actual finished articles. Otherwise, rubber compound is often regarded without proper respect for its value. Once this respect is established, good housekeeping will be perceived as of greater importance, and staff will be more careful to avoid contaminating compounds. And this respect must extend throughout management. One can hardly expect diligence from workers if management has ignored basic factors in maintaining compound integrity, such as construction of individual bays for separate mixing lines, and the physical separation of raw material from finished compound.

Engineers who are experienced in factory construction, but who are naive about mixing rubber compound, may save money by installing mixers in close proximity, without intervening walls, or by creating staging areas that combine raw material metering and delivery with accumulation of finished compound. But in the long run there are no such cost savings; they are consumed many times over by the cost of compound made useless by contamination. It is generally far easier to design a clean and efficient mixing plant than to try to enforce good procedures in a poorly designed facility.

5

Statistical process control for industrial mixing

R. J. Del Vecchio

5.1 INTRODUCTION

Rubber compounds have been mixed routinely for well over 150 years using a variety of types of equipment, most notably the two-roll mill and the internal mixer. These are batch processes which are sensitive in varying degrees to raw materials; there may be 5–20 raw materials in a given compound, all of which can vary appreciably in characteristics (especially polymers such as natural rubber). Other important factors are the mixing process itself, the type and condition of the equipment, the order of adding the ingredients and the time when each is added, the time and temperature exposure and the technique of the operator.

Given the large number of sources of variation in mixing, it is scarcely surprising that achieving a high degree of consistency over time among batches of any formulation has been difficult. When quality standards were comparatively low – when a successful mix produced a halfway decent molding and parts that could be recognized as rubbery – it was reasonable to tolerate a substantial variation in the mix.

In the context of the late twentieth century, following the wave of total quality [1,2] in both philosophy and practice, competitive companies have almost no choice but to achieve optimal quality and consistency in all processes. Every supplier of mixing equipment and raw materials will provide advice on how to get the best incorporation of ingredients, dozens of articles are written yearly on

The Mixing of Rubber
Edited by Richard F. Grossman
Published in 1997 by Chapman & Hall, London. ISBN 0 412 80490 5.

mixing procedures, and numerous types of computer software are available to help evaluate the output from the mixing process. Suppliers, mixers, formers and end users of rubber articles are all interested in knowing how the process is running and when to look for problems to solve or improvements to make.

For the past few decades the primary tool for fast and easy property checks of batches of mixed rubber has been the oscillating disk rheometer (ODR). Although newer and more sophisticated instruments have now become available, the ODR is still in wide general use and is likely to remain so for many years. This chapter will focus on the use of the ODR and the interpretation of its data through routine statistical process control (SPC) techniques.

Several software quality control (QC) programs are available for use with rheometer data, and some of them interface directly with the rheometer to generate run charts and process limits with minimal effort on the part of the user. This chapter considers the manual techniques, so the reader can choose whether or not to adapt them for use with software routines. A good understanding of manual procedures helps to reveal what the software does.

5.2 BASIC SPC CHARTING

Mixing is a batch or discrete process. Normal SPC charting for such processes employs what are known as X-bar and R charts, the data for which are generated by testing small subgroups (2–6 test pieces or batches) and averaging the resulting observations. At least 20 observations are preferred for good charting to begin. The ranges of all the averages for the initial observations are used to calculate an average range, known as $\bar{R}$.

$\bar{R}$ is actually an estimate of the normal scatter inherent in running a process even when that process is relatively stable. (Scatter is more technically known as variance, and the square root of variance, known as the standard deviation, is also commonly used to characterize the width of normal experimental observations.) By application of certain rules for interpreting the data, it is possible to make judgments with a high degree of confidence as to whether or not the process is stable.

Processes that are not stable are undergoing major changes while running, which makes it impossible to predict with any real accuracy what the process output will be at any instant. An unstable process must be investigated in order to determine what factors (assignable causes) are causing the major changes. The next step is to discover how to use appropriate factors in order to stabilize the process.

Only when a process has reached stability can valid judgments be made about the average properties of the process output and what ranges may be assigned to their normal variations. Typical SPC charts have lines that denote the usual limits of the stable process. The limits are calculated using the $\bar{X}$ and $\bar{R}$ numbers from an initial examination of the process; they may be recalculated at appropriate stages throughout the course of the process and as more data become available.

Examination of the data is easy in the context of the charts. Patterns can be found which show whether the process is running well, whether its output is occasionally untypical, or whether it is drifting over time. SPC charts can indicate when untypical output or product drift should be investigated and possibly corrected to restore process consistency.

When the process is stable, it is possible to determine how well the properties of its output match any given specification. If all its output falls within the callouts of the specification, the process is termed capable. The parameter CpK is one measure of how easily the process output fits within the specification callouts. A CpK of 1.0 indicates the process output just barely fits within the specification, whereas values such as 1.25 or 1.33 indicate there is at least some space between the outer limit of the process and the boundary of the specification.

These ideas flow from seminal work at Bell Laboratories during the early 1920s, work performed under the direction of Dr Walter Shewhart. They were used to good effect in some US industries before World War II and, following the war, they were popularized in Japan by Shewhart's students, most notably Dr W. Edwards Deming [3]. Other work was done by Amsden *et al.* [4]. and Grant and Leavenworth [5].

The underlying principle of using X-bar and R charts is that the variation within the subgroups is closely related to the overall process variation, and can be used accurately to calculate the process limits. In many discrete processes, such as a continuous stream of individual metal parts from some machining operation, this is valid and the technique works well. However, only a few rubber mixing operations, mainly those serving tire factories, mix long runs of one compound in the same equipment and thereby really fit the description of a classic discrete process. It is much more common for a variety of compounds to be mixed in any given day on some piece of equipment, with the variety sometimes covering very broad classes of polymer and ingredient types. Formulations may be run anywhere from one at a time to groups of batches numbering as many as 100 or more.

Because of this irregular type of production, use of standard X-bar and R charts does not always work well for rubber mixing. Instead there is a variation called X and mR charts. X means the actual observations are charted, not averages of subgroups; mR means a moving range is used to calculate $\bar{R}$ instead of the ranges of subgroups.

The most basic outline of the technique is as follows:

1. Run rheometer data on the compound of choice over time, until a reasonable number of batches has been made and tested (at least 12, preferably more than 20, and ideally over 30).
2. Enter the numerical data on the appropriate X and mR charts (Fig. 5.1) then put on the data points.
3. Calculate the moving range by determining the absolute difference between the observation for each batch and the observation for the previous batch. Enter it on the form.
4. Add up all the moving ranges (one fewer than the number of test observations) and divide by the number of moving ranges to get the average moving range ($\bar{R}$).
5. Calculate the upper control limit for the range (R_{UCL}) by multiplying $\bar{R}$ by 3.268, then draw this limit across the range chart.
6. Examine the chart to see whether any individual point falls above R_{UCL}; if so, the process is not in control and needs to be reviewed to discover and to lessen or eliminate the causes of variation before anything else can be done.
7. If the range chart is in control, proceed to calculate $\bar{X}$ by averaging all the individual X values.
8. Then calculate the upper and lower control limits (X_{UCL} and X_{LCL}) for the X chart using the following equations:

$$X_{UCL} = \bar{X} + 2.66\bar{R}$$

$$X_{LCL} = \bar{X} - 2.66\bar{R}$$

9. Draw the limits and also a line along the value of $\bar{X}$ onto the X chart and look for any of the following patterns which indicate a lack of control:
   - one point falling outside either limit
   - two points out of three on one side of the $\bar{X}$ line falling close to the control limit line
   - four points out of five on one side of the $\bar{X}$ line falling midway between the $\bar{X}$ line and the control limit line
   - seven points in a row falling on the same side of the $\bar{X}$ line
10. Any of the conditions in item 9 indicates that the process is not in control. The process should be reviewed to discover and to lessen or eliminate the causes of variation.

X and mR charts are not as powerful as X-bar and R charts in determining appropriate process limits, but they still serve well for many processes. In particular, if one or two points on the range chart barely exceed the R_{UCL} limit, it may not be altogether inappropriate to examine the X chart to see what it indicates. Should the X chart also give evidence

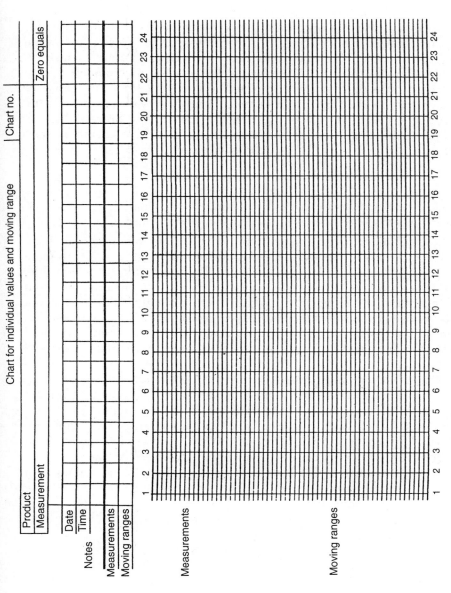

Figure 5.1 Enter the data on an X and mR chart.

of a control problem, appropriate action should be taken; but if it does not show any of the problem patterns in item 9, it may not be necessary to seek assignable causes of variation.

Evidence of an out-of-control process is not immediate justification for an operator adjustment; instead it should trigger an investigation of the process. Simple assignable causes may sometimes be found which do not themselves require any change to the process. A common example is a jump on the range chart related to changing from one group of batches to another group mixed later. Here the assignable cause is the change between the groups; it is not related to the process. Process adjustments should be approached with care and consideration; they should not be made as reflex actions in response to a less than perfect control chart.

5.3 RHEOMETER DATA AND ITS MEANING

The ODR generates an analog output of the torque required to oscillate a biconical disk immersed in the elastomeric compound exposed to some elevated temperature in a closed cavity. The compound starts off at the comparatively high viscosity characteristic of the material at room temperature. As the polymeric mixture becomes hotter from conduction, its viscosity drops towards its equilibrium viscosity at the cavity temperature. The measured torque therefore declines in the first minute or more after the test begins, then usually achieves a temporary plateau at some low level.

The heat eventually initiates the chemical reaction for curing or cross-linking of the polymer, and as the number of cross-links per unit volume increases, so does the apparent viscosity of the compound. The chemical kinetics of each compound's cross-linking gives it a signature curve in viscosity increase over time. When the cross-linking has been completed, or is very close to completion, the viscosity again plateaus out, but at a much higher level. The viscosity changes and stabilization are reflected in the torque requirements, which form a continuous curve when plotted versus time (Fig. 5.2).

The very varied chemistries of the polymers and cross-linking systems include some special cases. Some compounds never quite level off in viscosity, some reach a maximum then decline, and some have very steep or very shallow curves of torque over time. Rheometer curves are usually run at temperatures of 280–350 °F and for durations of 3–18 min. The temperature may sometimes be chosen to reflect the intended processing conditions, but this need not be the case. However, it is important to choose the time and temperature so as to allow full curing of the material.

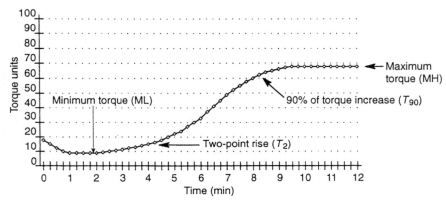

Figure 5.2 A typical rheometer chart.

There is no precise way to relate the data from the ODR curve to actual cured properties of the rubber, but some correlations can be made. The QC viewpoint is generally that consistency of the rheological response strongly implies consistency of related processing and final rubber properties.

Rather than compare the entire curve for each compound, it is customary to select certain parts of the information for simple inter-pretation and comparison. For instance, the minimum torque (M_{min}) measured in the initial part of the curve is called ML and reveals at least something about how well the material is liable to flow under processing conditions. Since the shear rate in the ODR is low compared to the shear rate in some common processes, especially injection molding, it is not safe to assume that the relative rating of compounds in ML is fully indicative of how they may behave under high shear.

Another important characteristic is the time required for the cross-linking to begin raising the compound viscosity, which often correlates with how much time there is for the compound to flow and form during processing. This is measured as the time required for the torque to increase from ML by a given number of units. The time intervals T_1 and T_2 are commonly used, the intervals required for the torque to rise by one and two units, respectively.

The maximum torque reached in the test (M_{max}) is called MH, and it correlates only roughly with the compound hardness or stiffness.

The time interval for the torque to achieve 90% of its increase from ML to MH is a measure of how long it takes for the compound to reach cure at the test temperature; it is known as T_{90}.

Other characteristics can also be obtained from the ODR curve, but the parameters described here are sufficient for most analyses.

5.4 A CASE HISTORY

A black reinforced natural rubber (NR) compound had been in long usage at a molding house. During the course of implementing total quality management (TQM) the company targeted the compound to improve its quality and consistency. The material was not used in great volume, and was mixed in a #3 Banbury no more often than several times per month. As few as one and as many as four batches might be mixed at any given time. Rheometer data had been generated for years, and all the data from the start of one year through the following October were used as the initial rheological data set (Table 5.1).

The appropriate charts for mR and X of several properties were constructed and immediately demonstrated that the process was out of control. Careful examination of the charts indicated that discernible process shifts could be found in the course of the year, along with a high degree of scatter. For the sake of brevity only two properties will be used to illustrate the situation, although Table 5.1 contains seven rheological properties and the durometer measurement, too.

The two properties are ML and T_2. ML is strongly related to the basics of the polymer, reinforcement system and process aids; T_2 is more dependent on the chemistry of the cross-linking system (also known as the curative package). With this particular NR compound, these two properties will definitely be sensitive to changes in polymer, ingredients and processing. Table 5.2 demonstrates the numbers used to generate the limits of the ML chart, and Figs 5.3 through 5.6 are the actual mR and X charts for ML and T_2.

Figure 5.3 shows how two points exceed R_{UCL} just as noted in Table 5.2, but the discrepancies are much easier to find using the chart. The upper chart limit is almost at 5 torque units, which is a large span for minimum viscosities; for a single compound the range of ML from batch to batch would often be less than 3 torque units. Thus the chart indicates an out-of-control condition, even though a generous limit was used. This reflects badly on the process consistency, but it is still possible to plot and examine the X chart.

Figure 5.4 immediately shows the same points beyond the lower limit as can be found in Table 5.2; moreover, it clearly shows a long run on one side of the centerline (the 11th point through the 22nd point) which also indicates a process out of control. The overall pattern of the points visually demonstrates wide variation and the limits are almost 8 torque units apart, again very wide for batches of a single formulation.

The mR chart for T_2 (Fig. 5.5) also shows a point beyond R_{UCL} and a generous upper limit of about 1.4 min. The X chart for the property (Fig 5.6) has no points outside either limit, but starts off with a run of

Table 5.1 Rheometer data for case history of section 5.4

Date	ML	Time to ML	T_2	T_{90}	Cure rate	MH	Time to MH	Duro
91 01 16	9.73	2.67	4.50	7.97	11.40	38.36	8.73	67
91 02 20	7.21	2.10	3.90	7.40	16.80	41.95	8.10	63
91 02 20	8.31	1.90	3.80	7.30	16.80	43.25	8.00	63
91 02 21	7.48	1.98	3.98	7.42	13.20	38.31	8.07	61
91 04 12	12.08	2.22	3.97	6.93	15.00	40.50	7.53	66
91 05 03	11.55	2.23	4.03	7.05	13.20	38.90	7.60	62
91 05 03	12.50	1.88	3.60	6.50	12.60	38.95	7.07	63
91 05 04	11.05	1.55	3.50	6.95	12.00	38.90	7.68	66
91 05 04	11.75	1.88	3.45	6.03	18.00	42.80	6.52	66
91 05 04	11.55	2.17	4.28	7.92	7.20	32.30	8.78	64
91 06 18	13.62	1.73	3.70	6.37	15.60	41.55	6.87	66
91 07 25	14.65	1.78	3.02	5.07	24.00	49.00	5.52	61
91 07 25	14.15	1.60	2.85	4.93	21.60	49.75	5.38	62
91 07 25	14.55	1.95	3.32	5.45	20.40	45.40	5.88	56
91 07 25	14.38	1.63	3.20	5.47	21.00	45.94	6.00	57
91 07 26	12.10	1.35	2.48	4.52	23.40	45.00	5.02	60
91 07 26	12.25	1.47	2.50	4.42	25.20	46.20	4.85	60
91 07 26	12.70	1.50	3.65	6.20	16.80	42.15	6.62	68
91 07 26	12.30	1.38	2.95	5.22	20.40	43.60	5.68	68
91 07 27	13.00	2.13	3.98	6.68	12.60	39.15	7.20	66
91 07 27	12.95	1.30	2.62	4.58	21.60	44.50	4.98	67
91 07 31	13.30	1.52	3.58	6.13	15.00	40.64	6.62	68
91 07 31	11.58	1.13	2.80	5.48	14.40	38.33	6.08	67
91 08 05	13.90	1.13	2.42	4.62	25.80	51.12	5.47	60
91 08 20	12.67	1.73	3.73	6.27	17.40	41.20	6.92	65
91 08 22	11.44	1.88	3.95	6.65	15.00	39.05	7.23	65
91 08 26	11.19	1.50	4.05	7.12	12.60	38.60	7.83	66
91 09 26	13.65	1.80	3.68	6.43	15.00	44.56	7.02	70
91 09 26	14.25	2.05	4.17	7.53	9.00	37.89	8.28	68
91 10 04	12.34	1.25	2.70	4.88	21.60	49.13	5.18	69
91 10 04	9.38	1.30	2.63	4.67	25.80	47.82	4.95	66
91 10 04	9.03	1.48	2.93	5.07	21.60	44.46	5.38	68
91 10 07	12.29	1.67	3.18	4.92	33.60	51.34	5.13	68
91 10 07	8.68	1.50	2.95	4.62	36.00	46.52	4.82	69
91 10 15	12.82	2.10	4.02	6.85	12.60	38.46	7.48	70
91 10 15	13.47	2.03	3.70	6.20	16.80	42.95	6.70	70
91 10 15	12.47	1.62	3.63	6.03	16.20	41.55	6.53	70
91 10 16	8.10	1.82	4.07	6.63	11.40	30.33	7.18	62
91 10 16	8.35	1.25	3.32	5.43	18.00	35.51	5.85	63
91 10 16	7.95	1.42	3.50	5.67	17.40	33.60	6.08	62
91 10 16	13.33	1.87	3.60	6.00	17.40	43.61	6.48	62
91 10 16	13.18	1.92	3.45	5.75	20.40	44.12	6.25	62
91 10 16	13.08	1.73	3.52	5.78	18.00	42.05	6.28	60
91 10 16	12.58	1.17	2.90	5.57	18.60	43.66	6.15	63

Table 5.2 Calculation of X and mR limits for rheometer ML of 44 batches

Batch	1	2	3	4	5	6	7	8	9	10	11	12	13	14	15
ML	9.73	7.21	8.31	7.48	12.08	11.55	12.50	11.05	11.75	11.55	13.62	14.65	14.15	14.55	14.38
mR		2.52	1.10	0.83	4.60	0.53	0.95	1.45	0.70	0.20	2.07	1.03	0.50	0.40	0.17

Batch	16	17	18	19	20	21	22	23	24	25	26	27	28	29	30
ML	12.10	12.25	12.70	12.30	13.00	12.95	13.30	11.58	13.90	12.67	11.44	11.19	13.65	14.25	12.34
mR	2.28	0.15	0.45	0.40	0.70	0.05	0.35	1.72	2.32	1.23	1.23	0.25	2.46	0.60	1.91

Batch	31	32	33	34	35	36	37	38	39	40	41	42	43	44
ML	9.38	9.03	12.29	8.68	12.82	13.47	12.47	8.10	8.35	7.95	13.33	13.18	13.08	12.58
mR	2.96	0.35	3.26	3.61	4.14	0.65	1.00	4.37	0.25	0.40	5.38	0.15	0.10	0.50

$\bar{R}$ (average all mR) = 1.40

$R_{UCL}(3.268\bar{R})$ = 4.58

$\bar{X}$(average of all X) = 11.79

$X_{UCL}(\bar{X} + 2.66\bar{R})$ = 15.52

$X_{LCL}(\bar{X} - 2.66\bar{R})$ = 8.06

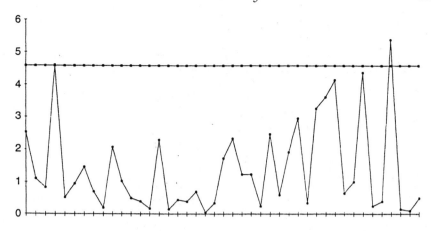

Figure 5.3 mR chart for ML: Jan 91 to Oct 91, 44 batches.

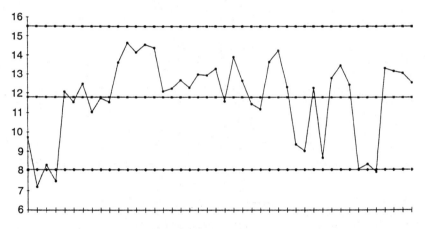

Figure 5.4 X chart for ML: Jan 91 to Oct 91, 44 batches.

11 points on one side of the centerline, which demonstrates lack of control. The limits are about 2.5 min across, again a fairly broad distribution.

These data show there is no question that the process was running with less than adequate consistency over the 10 months by the criteria of these two properties. Some of the other properties in Table 5.1 can be charted as an exercise by the reader, and will also demonstrate lack of process control.

Once the undesirable state of the process had been established, meetings were held and possible causes of the excessive variability were

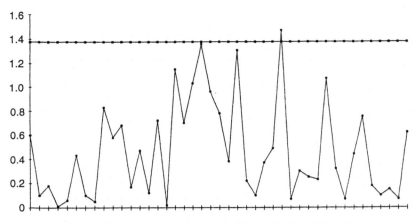

Figure 5.5 mR chart for T_2: Jan 91 to Oct 91, 44 batches.

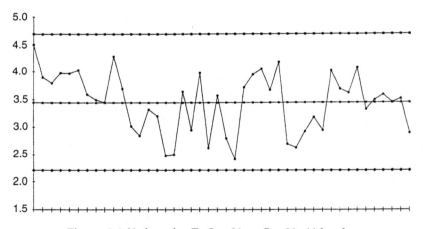

Figure 5.6 X chart for T_2: Jan 91 to Oct 91, 44 batches.

discussed. The mixing procedure had not been updated for some years and operators varied in the amount of time the NR was peptized before additions were made, as well as the order of ingredient additions and the temperatures used during the cycle. As a first step, the procedure was revised slightly and all operators instructed in its use. By December new mix sheets were available at the Banbury and everyone was familiar with the standard mixing cycle.

The mR chart (Fig. 5.7) for the next four months has no points beyond or even very close to the control limit. And that limit is just about 2.5 torque units, largely where other experience indicated it should be.

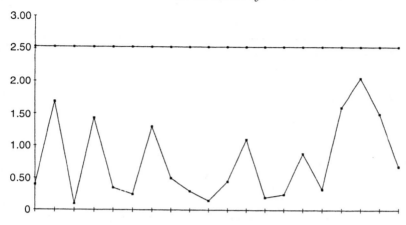

Figure 5.7 mR chart for ML: Dec 91 to Mar 92, 21 batches.

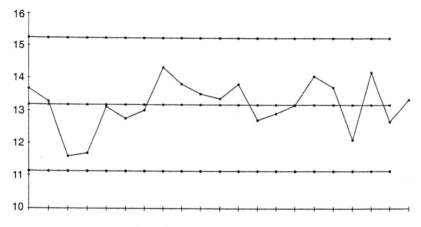

Figure 5.8 X chart for ML: Dec 91 to Mar 92, 21 batches.

The X chart in Fig. 5.8 reveals a pattern of points demonstrating good control and much narrower limits than had previously applied. The charts for T_2 (Figs 5.9 and 5.10) are likewise in control with limits reduced from those of the previous 44 batches. The remedial measures taken to improve the mixing procedure had clearly been effective. A reduction was noted in troubleshooting levels for molded parts requiring this compound, and none of the special batch dispositions or reworks, which had to be used up until then, was employed once the new procedures had gone into effect.

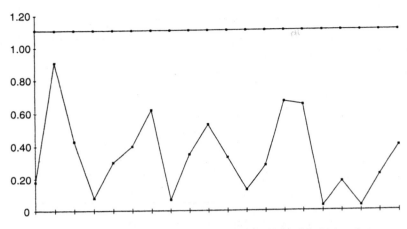

Figure 5.9 mR chart for T_2: Dec 91 to Mar 92, 21 batches.

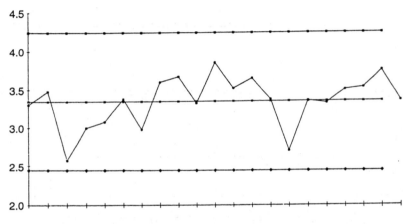

Figure 5.10 X chart for T_2: Dec 91 to Mar 92, 21 batches.

Subsequently it was decided to attempt further improvements and other changes to the process. The principal difference was to check weighing of all ingredients, with operators signing off on the daily process sheets. During the following three months the overall batch-to-batch variability in ML decreased further, as shown by the lower R_{UCL} of the mR chart (Fig. 5.11); but there was a shift and return in the individual measurements, as indicated by the pattern in the X chart (Fig. 5.12). Variability in T_2 increased (Fig. 5.13) but the property stayed in control with an almost classic random pattern (Fig. 5.14).

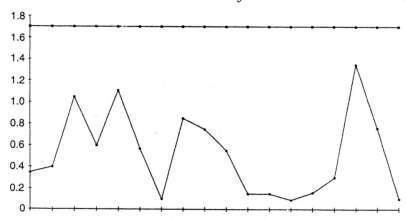

Figure 5.11 mR chart for ML: Apr 92 to Jun 92, 19 batches.

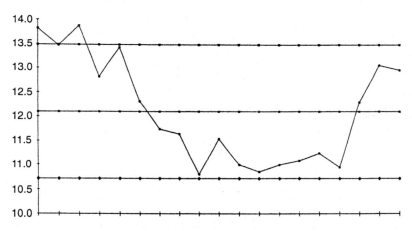

Figure 5.12 X chart for ML: Apr 92 to Jun 92, 19 batches.

Investigation led to some fine tuning of the formulation itself, including minor changes in ingredient sources and levels. Data for the last four months of the project are shown in Figs 5.15 to 5.18. The level of variation in ML has been reduced further, but the minimum viscosity of one batch out of the 27 did fall below the process limit. The process was checked at that time and the assignable cause was probably 'generous' weighing of processing oil to that batch and the previous two batches by a new operator. The extra oil was not sufficient to affect other properties significantly, but it did cause the batch viscosities to drop just enough to be noticeable.

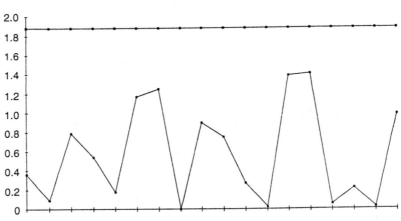

Figure 5.13 mR chart for T_2: Apr 92 to Jun 92, 19 batches.

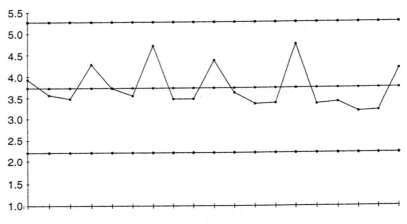

Figure 5.14 X chart for T_2: Apr 92 to Jun 92, 19 batches.

T_2 variation also lessened, and the X chart had no points outside the limits. However, there appeared to be a process shift beginning at about the ninth batch, since the next seven batches fell on the same side of the $\bar{X}$ line. In this case no action was taken since the seventh batch was at the beginning of a four-batch series, the rest of which showed an increase in T_2 back up towards and over the centerline. There may have been an undetected transitory cause of the property drift, or it may have been due to chance.

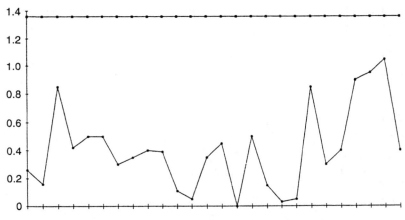

Figure 5.15 mR chart for ML: Jul 92 to Oct 92, 27 batches.

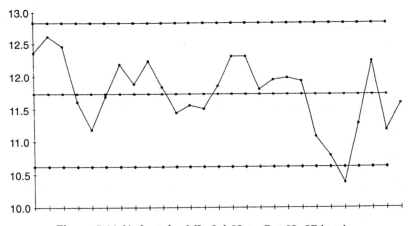

Figure 5.16 X chart for ML: Jul 92 to Oct 92, 27 batches.

There is a huge contrast between the original process and its third improved version. The original process was not stable, a serious concern to begin with. But even neglecting that, although the average level of ML did not change significantly (less than 1%), scatter as measured by the standard deviation went from 2.1 to 0.5 torque units, a 75% reduction. T_2 also remained essentially unchanged, but its standard deviation dropped from 0.54 to 0.28 min, an almost 50% reduction. These changes would dramatically raise CpK if the same specification were applied to the old and new processes.

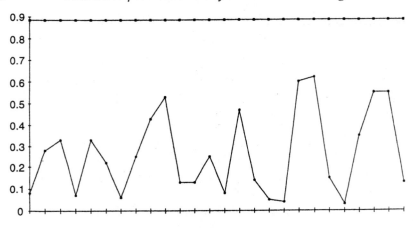

Figure 5.17 mR chart for T_2: Jul 92 to Oct 92, 27 batches.

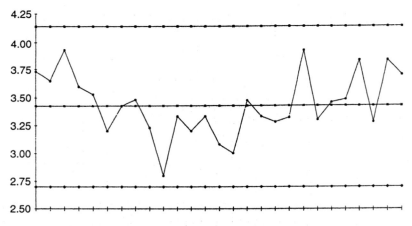

Figure 5.18 X chart for T_2: Jul 92 to Oct 92, 27 batches.

5.5 SUMMARY

Overall the charts show that SPC is a convenient and efficient technique for investigating how well a process is operating. They show how problems can be found and how effective problem solving can help to develop a more consistent process then refine it into a more capable process. This was the thrust of Shewhart's approach: start by using SPC to get a process under control; find and eliminate assignable causes of sporadic variation. Continue by using it to find and minimize the sources of normal or chronic variation in the process.

SPC need not be limited to rheometer data; physical properties, such as durometer, tensile strength, elongation and tensile modulus, could be analyzed the same way, as could other more specialized rubber properties, e.g. Mooney viscosity or scorch. All of them will be affected by the mixing process, although some of the physical properties are inherently more widely distributed than rheological data, and less meaningful in regard to the rubber's capacity for being processed in steps such as molding, extruding or calendering. Rheometer data were described in this chapter because they are currently the most common quality measurements made on mixed batches in the rubber industry.

5.6 MIXING CONTROL SOFTWARE

Several software packages are available which will interface directly with the rheometer and painlessly perform this type of data compilation and analysis. Various options may be built into the software which change aspects of the analysis; for instance, subgroups could be used instead of individual data points, or the thresholds used to indicate process deviations might be altered to have higher or lower sensitivity to possible deviations.

Even though selected nonrheological properties may be entered separately, they may still be analyzed by the software. It is important to choose properties that directly affect the performance of the material or product; and if they have no direct effect, there should at least be a close correlation.

Such programs need to be user-friendly and provide full access to all the data as they accumulate. The type and clarity of the reports available from the software are of key importance to TQM efforts, and potentially to the customers. It should at least be possible to examine the controlling equations for the charts, perhaps even access them or select alternatives. The system might be customized, if necessary, depending on the default settings and what is known of the process characteristics.

No matter how easy a program is to use, its output and function should occasionally be reviewed by someone conversant with quality and SPC techniques. Computers cannot really do what people want them to; they can only do what they are told, and it is very important that the company has within it the ability to examine the procedures applied to the data and to judge how well they fit each particular operation.

REFERENCES

1. Feigenbaum, A. V. (1961) *Total Quality Control*, McGraw-Hill, New York.
2. Ishikawa, K. (1985) *What is Total Quality Control?* Prentice Hall, Englewood Cliffs NJ.
3. Deming, W. E. (1982) *Quality, Productivity, and Competitive Position*, MIT Press, Cambridge MA.
4. Amsden, R. T., Butler, H. E., Amsden, D. M. (1989) *SPC Simplified*, Quality Resources, New York.
5. Grant, E. L. and Leavenworth, R. S. (1988) *Statistical Quality Control*, McGraw-Hill, New York.

6

Additives that affect mixing

Robert F. Ohm

6.1 INTRODUCTION

Most additives that affect mix procedures are reactive chemicals. To run a chemical reaction successfully in a Banbury internal mixer or using other equipment, one needs to control three variables: time, temperature and stoichiometry (the ratio of the reactants). This chapter will discuss the effect of these variables on the chemical peptizing of natural rubber (NR), styrene–butadiene rubber (SBR), polychloroprene (CR) and Thiokol polysulfide. Other topics will include additives to increase viscosity, when to add zinc oxide, filler treatments and bin storage problems.

6.2 PEPTIZERS IN NATURAL RUBBER

From the earliest days rubber technologists have faced the problem of reducing the viscosity of rubber to make it processable. In the 1820s Thomas Hancock used a roller with grooves or pins, which he called a pickle, to break down the very high molecular weight of natural rubber. This mastication process reduced the viscosity, as collected from the tree, to a more usable viscosity for making macintosh raincoats and other rubber articles. Reducing viscosity strictly by mechanical means has an inherent limitation. When the rubber is cold, the mixer provides high shear and polymer breakdown occurs rapidly. However, mixing creates heat. As the mix warms, thermal softening provides less shear and the breakdown of molecular weight becomes slower. Chemical additives called **peptizers** have been developed to hasten this molecular cleavage.

The Mixing of Rubber
Edited by Richard F. Grossman
Published in 1997 by Chapman & Hall, London. ISBN 0 412 80490 5.

Additives that affect mixing

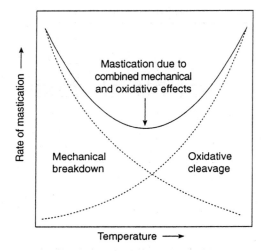

Figure 6.1 Effect of temperature on mastication.

As with all chemical reactions, the reaction of peptizers with NR occurs more rapidly as the temperature is increased.

6.2.1 Effects of temperature

With the use of chemical peptizers, one generally sees a U-shaped curve of activity versus temperature (Fig. 6.1). At the lowest temperature, the rate of breakdown is rapid because of high mechanical shear. As the temperature increases, the rate of breakdown declines to a minimum. At still higher temperatures, the chemical activity of the additive becomes the dominant effect and the rate of breakdown increases again. The minimum generally occurs at or slightly above 100 °C. As shown in Fig. 6.1 it is generally desirable to keep the breakdown as cool as possible for maximum breakdown and productivity when mixing on a mill. Conversely, in internal mixers it is most productive to break down under warm or hot conditions, but very careful control of mix temperature is required to obtain uniform results from batch to batch. Typical data with various mixing conditions are shown in Table 6.1 [1].

6.2.2 Effects of time

Adding a chemical peptizer will allow the desired viscosity to be achieved in a shorter time, as shown in Fig. 6.2. Note also that the use of a peptizer on an open mill provides more uniform mastication, which

Table 6.1 Effect of mix temperature on Mooney viscosity [1]

Temperature at start	Mooney viscosity ML (100 °C) 1 + 4		
	No peptizer	0.25 phr peptizer	0.5 phr peptizer
Open mill			
Mix time = 10 min			
70 °C	56	40	38
100 °C	74	40	39
Internal mixer			
Mix time = 4 min			
150 °C	94	53	44
160 °C	–	49	38

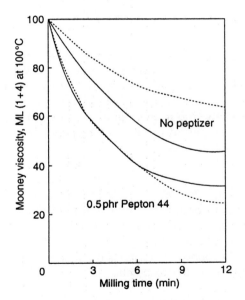

Figure 6.2 Effect of milling time on natural rubber: (- - -) at 100°C and (——) at 70°C.

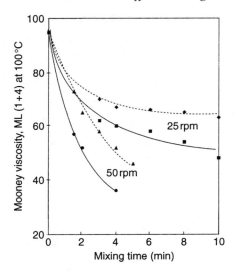

Figure 6.3 Effect of time and speed: (- - -) no peptizer and (———) 0.1 phr Renacit VII.

is less sensitive to changes in mix temperature. In contrast, internal mixers are very sensitive to changes in mixing time or rotor speed, as shown in Fig. 6.3 [2]. High speed mixing reduces viscosity quite rapidly, but the batches may dump in a very sticky and nonuniform state. This problem may be partly due to inadequate dispersion of the peptizer. To maximize dispersion, the addition of extra plasticizers or process aids has also been suggested [3].

6.2.3 Effects of use level

The effects of different dosages of two common peptizers are shown in Fig. 6.4 [2]. Within the range commonly used there is no robust dosage that is insensitive to changes in peptizer level. Therefore, it is crucial to control carefully the peptizer dosage and dispersion to obtain uniformity in viscosity from batch to batch.

6.2.4 Effects of other additives

Peptizers for natural rubber are mixtures of materials that hasten oxidation and stabilize the broken polymer chains. Thus, it is counterproductive to add antioxidants during the mastication step. Data to show the effects of several additives on mastication are given in Table 6.2. Some additives completely inhibit peptizer activity; this prevents a

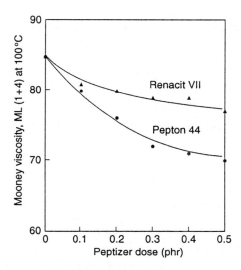

Figure 6.4 Effect of peptizer concentration.

Table 6.2 Effect of compounding ingredients on peptizing efficiency [1]

	Mooney viscosity ML (100 °C) 1 + 4	
	No peptizer	*0.25 phr Pepton 44*
NR mixed 15 min at 90 °C initial to ~115 °C final		
No additive	76	41
With inhibitors		
Sulfur	81	64
Vanox MBPC	80	59
Wingstay 100	79	59
Methyl Tuads	76	58
Altax	75	51
Stearic acid	77	47
With facilitators		
Vanax DPG	62	43
Zinc stearate	69	39

Table 6.3 Use of peptizer in SBR 1712[a] [1]

Peptizer dose (phr)	0	1.0	2.0	3.0
ML (100 °C) 1 + 4	40	36	33	31

[a]All mixes 4 min at 160 °C. SBR 1712 has a nominal Mooney viscosity of 55.

vulcanized rubber article from turning to soup in a dynamic service environment.

6.3 PEPTIZERS IN SBR

The use of chemical peptizers is generally much less effective in unsaturated synthetic rubbers than in natural rubber. This is partly because the structure of natural rubber makes it more susceptible to cleavage, compared to butadiene-based polymers, which are resistant to molecular weight breakdown. It also stems from the use of antioxidants as stabilizers in synthetic polymers. Nevertheless, chemical peptizers can be used in selected synthetic rubber applications where Mooney viscosity reduction is desired. For example, sponge rubber compounds benefit from lower viscosity to produce lower density sponge.

Typical data is given in Table 6.3; the mix time was 4 min at 160 °C. SBR 1712, as received, has a nominal Mooney viscosity of 55.

In many cases the use of plasticizers and processing aids will be equally effective in reducing viscosity. This approach has the advantage of being less sensitive to mix procedures. Examples of plasticizers and processing aids for an SBR 1500 compound are shown in Table 6.4. At higher levels their use will reduce the hardness and modulus [4].

6.4 PEPTIZERS IN SULFUR-CONTAINING POLYMERS

Some grades of neoprene polychloroprene and Thiokol polysulfide can also be reduced in viscosity by mechanical or chemical action. This is due to the sulfur–sulfur bonds in the polymer backbone, which can be easily cleaved. Neoprene G-types are peptizable by the use of additives such as Altax, Vanax DPG or Vanax 552. Vanax 552 is particularly active, as shown in Fig. 6.5 and as observed in natural rubber, the viscosity obtained with a peptizer is less sensitive to temperature variation than breakdowns without a peptizer. Various peptizers for neoprene are compared in Table 6.5. The W-type neoprenes do not have sulfur–sulfur bonds and, unlike the G-type neoprenes, they are not peptizable or subject to breakdown under mechanical shear [5].

Thiokol FA polysulfides respond in a similar manner to neoprene when Altax is used to break sulfur–sulfur bonds. The data is given in

Table 6.4 Plasticizers and process aids in SBR [4]

	No additive	*With Vanplast R*	*With Vanplast PL*	*With Leegen*	*With Vanfre AP-2*
ML (100 °C) 1 + 4	85	75	72	76	73
Cured 15 min at 153 °C					
200% modulus (MPa)	12.3	10.8	10.7	9.1	10.5
Tensile strength (MPa)	14.1	14.0	14.1	12.7	14.0
Elongation (%)	260	290	320	370	290
Hardness (Shore A)	70	70	68	68	70
DuPont Flow Mold: weight of Spider (g)	12.5	13.8	13.4	13.4	14.8

Base compound	
Plioflex 1500C	100
Stearic acid	2
Zinc oxide	5
Agerite Stalite S	1.5
Thermax (N-990)	120
HAF black (N-330)	10
Sulfur	2
AMAX	1
CUMATE	0.15
Additive (optional)	5

Table 6.5 Peptization of neoprene GNA [5]

Sample[a]	*MS (100 °C) 1 + 2.5*
No peptizer	50
With Altax	43
With Vanax DPG	41
With Vanax 552	12

[a]All samples mixed 10 min at 50 °C (121 °F).

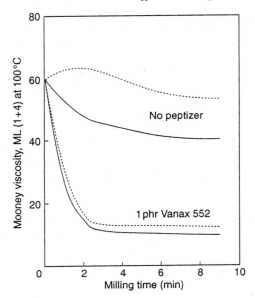

Figure 6.5 Peptization of neoprene GN: (-----) at 70°C and (———) at 40°C.

Table 6.6 Peptization of Thiokol FA [6]

Altax (phr)	0	0.25	0.30	0.35	0.40
ML (100 °C) 1 + 4	96	76	66	57	43

Table 6.6; all stocks with Altax also contain 0.1 phr Vanax DPG to activate the Altax. Thiokol ST polymers are polymerized to a desired Mooney viscosity and are rarely peptized [6].

6.5 ADDITIVES TO INCREASE VISCOSITY

Some polymers, notably butyl, have a tendency to cold flow. And low hardness compounds use copious quantities of oils, which also reduce viscosity and cause cold flow problems. This characteristic of cold flow can be particularly annoying after preforms have been made. Chemical additives such as Vanax PY can increase the viscosity of the butyl mix and reduce the tendency to cold flow. Heat is required for this chemical reaction, so the polymer should be mixed hot. Zinc oxide and some of the carbon black can be added at the initial step. However, accelera-

Table 6.7 Vanax PY in butyl [7]

Vanax PY (phr)	0	0.2	0.4
ML (100 °C) 1 + 4	89	87	99
Cured 60 min at 153 °C			
300% modulus (MPa)	8.6	11.2	12.8
Tensile strength (MPa)	15.0	16.9	15.5
Elongation (%)	500	400	380
Hardness (Shore A)	65	61	60
Yerzley resilience (%)	45	51	52
Heat buildup: Goodrich temperature (°C)	63	56	54

	Base compound	
	Exxon butyl 217	100
	N-330 carbon black	15
	S-315 carbon black	35
	Stearic acid	1
	Zinc oxide	5
	Altax	1
	Methyl Tuads	1
	Sulfur	2

tors should be withheld from the first-pass mix and added on the mill or in a cooler second pass. Data on the use of Vanax PY is shown in Table 6.7. The modulus is significantly higher and the resilience is also increased for compounds containing Vanax PY [7]. (Note that Yerzley resilience does not correlate well with Bashore or other rebound resilience methods, since a butyl ball will bounce less than 10%.)

The viscosity can also be built up by physical means. An example would be Factice vulcanized vegetable oil. The sulfur cross-linking of vegetable oils produces a solid which flows under shear, not under heat. These additives are very useful in building viscosity of low hardness compounds that use large quantities of hydrocarbon oils or ester plasticizers. Incorporation of Factice helps to prevent the oil from bleeding or exceeding its compatibility limit in the rubber, as well as maintaining a workable viscosity. Compounds should be mixed by first adding the Factice. Adjust the mill for a tight setting, allow the Factice to fall through and leave the crumbs in the mill pan. Then, after banding the polymer, the crumb is added in the first sweepings. The powdered grades of Factice do not need such critical timing when they are added to the mix.

6.6 PREVENTING UNWANTED CHEMICAL REACTIONS

Zinc oxide, an ingredient in most formulations, is a dense material that tends to compact and disperse with difficulty. For optimum dispersion it is most commonly added early in the mix. However, there are some systems where late addition of zinc oxide is beneficial. Neoprene is rapidly cross-linked by zinc oxide, so ZnO is typically the last ingredient added to a neoprene mix. To increase scorch protection, magnesium oxide and Agerite Stalite S are added early in the mix, generally the first items after the polymer is banded. In neoprene, polymerized quinoline antioxidants and most types of p-phenylene antioxidants cause scorchy compounds, so they are not used. For ozone protection, Wingstay 100-AZ is the least scorchy antiozonant in neoprene. Carboxylated NBR rapidly undergoes ionic cross-linking on addition of zinc oxide; again, late addition of ZnO is recommended. To ensure a homogeneous mixture with late addition, predispersed forms of zinc oxide can often be beneficial. Moisture can also exacerbate ionic precure problems, so it is important that all ingredients are bone dry. Fillers with a high surface area (such as precipitated silica) are a common source of moisture, particularly if using broken or partial containers. Incipient

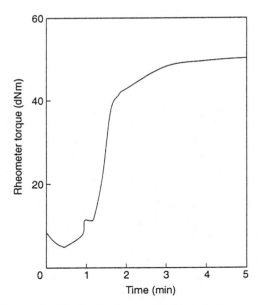

Figure 6.6 Partial ionic cross-linking during cure.

scorch due to ionic cross-link formation can sometimes be seen as a hump in the ODR curve after T_{s_1} or T_{s_2}, as shown in Fig. 6.6 [8]. Phenolic antioxidants are generally the only sort recommended for carboxylated NBR; this is because amines compete with zinc oxide for the carboxyl groups. Neoprene and carboxylated nitrile blends must be mixed with especial care because they are self-vulcanizing [9].

Zinc oxide causes many polymers to rapidly lose HCl (or HF) and degrade. Zinc oxide is therefore avoided in Hypalon, chlorinated polyethylene, polyacrylate, Vamac, epichlorohydrin, Viton, Aflas and fluorosilicone elastomers.

6.6.1 Filler treatments

In order to attain optimum properties, some fillers can benefit from chemical treatments. To properly control this reaction, attention to the mixing order is required. Hi-Sil silica is a very active filler which will absorb and deactivate accelerators and zinc oxide. So when compounds contain Hi-Sil, the accelerators and zinc oxide are often added last, after the silica surface has been passivated with a polyhydroxy compound such as Carbowax polyethylene glycol, diethylene glycol, glycerine or triethanolamine. An example is shown in Table 6.8. The use of triethanolamine will tend to cause scorchy compounds.

Silane treatment of mineral fillers can increase modulus, reduce compression set and minimize water swell. The mineral filler must have hydroxyl groups on the surface to react with the silane and bind it to the mineral surface. The silane should therefore be added shortly after the filler becomes incorporated, and before other reactive chemicals are added. Some mineral fillers may be purchased already blended with silanes to avoid this complication in mixing.

Wire adhesion stocks containing reactive resins are used to make steel-belted tires. These compounds require particular care in mixing to achieve optimum adhesion. Resorcinol is typically added in the first pass so it will melt and disperse. Resorcinol–formaldehyde resins can be substituted to minimize fuming. The methylene donor which reacts with the

Table 6.8 Effect of Carbowax in compounds containing Hi-Sil silica [10]

Carbowax 4000 (phr)	0	2	4
ML (100 °C) 1 + 4	51	41	39
Mooney scorch, T_5 at 135 °C (min)	>30	12	9
Optimum cure at 144 °C, $T_{c_{90}}$ (min)	34	19	16

resorcinol, either hexamethylene tetramine (hexa) or hexamethoxymethyl melamine, goes in the second pass. Hexa is scorchy and, because it is a solid, disperses with more difficulty than the liquid melamine. Insoluble sulfur is used in wire adhesion stocks to prevent sulfur bloom at ≥ 3 phr dosages typically used. High dosages are needed for maximum adhesion to the brass-coated wire surface. Insoluble sulfur reverts to the ordinary rhombic form above 100 °C, so it must be added in the second, cooler pass to keep it nonblooming.

6.6.2 Bin storage problems

In some operations a mixed compound will be used over several days or even weeks. Ideally the cure characteristics would not change over this period of time. However, we do not live in an ideal world and must cope with bin storage problems. Many types of compounds can exhibit poor bin storage. A few examples are ultrafast cures for extrusions vulcanized at atmospheric pressure, aldehyde–amine cures of chlorinated polyethylene and some peroxide cures of EPDM. One technique to consider with bin storage problems is acceleration of the mix on the day it is used.

REFERENCES

1. American Cyanamid Company (1972) Pepton 44 Plasticizer for the Mastication of Natural Rubber and Synthetic Polymers.
2. Crowther, B. G. (1981) Peptization of natural rubber in an internal mixer. *NR Technology*, **12**(2), 27–35.
3. Crowther, B. G. (1983) Peptization of natural rubber in an internal mixer 2. Fatty acid soaps and blends with chemical peptizers. *NR Technology*, **14**(1), 1–7.
4. Ohm, R. F. (ed.) (1990) *The Vanderbilt Rubber Handbook*, R. T. Vanderbilt Norwalk CT, p. 454.
5. Schmitt, S. W. (1965) Peptizing agents for neoprene, *DuPont Technical Literature NP-360.1*.
6. Hanson, C. (1992) Personal communication, Morton International, Woodstock Research Center.
7. DuPont (1956) Polyac Chemical Conditioner for Butyl Rubber, *DuPont Bulletin 22*.
8. Goodyear Chemicals (1976) CHEMIGUM NX 775 Scorch Resistant Carboxylated Nitrile Rubber.
9. Mukhopadhyay, S. and De, S. K. (1992) *J. Appl. Polym. Sci.*, **45**, 181–85.
10. Hewitt, N. L. (1978) Reinforcing NR and polyisoprene with silicas. *Eur. Rubber J.*, **1978** 11–19, 73.

7

Operation and maintenance of mixing equipment

Steven R. Salma

7.1 INSPECTION OF BANBURY MIXERS

The Banbury mixer is a rugged piece of equipment, manufactured to exacting specifications and tolerances, intended for maximum life and productivity. Service life will range from 2 years (for extremely abrasive compositions) up to 25 years (for soft compounds such as sponge rubber). As the mixer ages, the tips of the wings of the rotor become rounded, reducing the shear rate. In later stages, the hard surfacing of the chamber sides begins to wear, further reducing the shear rate as well as increasing the fill factor. Wear to the throat increases material sticking, and excess clearance reduces pressure on the compound. As overall wear increases, the batch size is usually increased and the cycle time is often extended. Invariably the quality of dispersion becomes adversely affected. The end of useful life is best determined by monitoring the quality of the compounds and production rates. Once quality starts to fall off, it will drop at an increasing rate because the wear resistance of the hard surface layer diminishes as the base metal is approached. When product quality first starts to drop, plans should be made for replacing the mixing chamber.

Scheduled preventive maintenance is vital to ensure operation with minimum downtime and expense, and maximum service life. And any variables that affect process control (rotor speed, rotor, body and door temperature, coolant flow rate and ram pressure) must be checked to make sure they correspond to instrumentation readings.

The Mixing of Rubber
Edited by Richard F. Grossman
Published in 1997 by Chapman & Hall, London. ISBN 0 412 80490 5.

Determination of the overall mechanical condition of the mixer, and estimation of when component replacement will be required, necessitate formal inspection at regular intervals (typically by the manufacturer's technologists). The objectives are as follows:

- To determine whether process control instrumentation remains accurate.
- To determine whether the lubrication program is adequate.
- To determine whether machine adjustments are indicated.
- To determine whether any components should be replaced.

The following components are inspected:

- rotors and bearings
- bearing and mixer lubrication systems
- side, rotor and doortop cooling systems
- dust stops and related lubrication
- drop door and latch
- hydraulic system
- gears and couplings
- Banbury hopper

Before starting the inspection, it is essential to complete the shutdown procedure specified in the instruction manual. Restarting for test purposes should be done under engineering supervision.

7.1.1 Inspection at the mezzanine level

Side cooling

The outside wall should be at the same temperature at both top and bottom zones; if there is a differential, a zone may be partially plugged. If flow rates are correct, the outlet pipes should run 6–10 °F higher than the inlets.

Rotor cooling

Like the side cooling, a 6–10 °F temperature rise is expected. If higher, check the flow rate. If very low, check the siphon pipes (if broken, high rotor temperatures and almost no temperature rise is found).

Rotors and bearings

After removal of grease and traces of compound from rotors in the dust stop sections, the mixer is started and rotor movement observed. Random vertical movement is an indication of bad bearings, bearing bores or sleeves. Vertical rotor movement in time with rotation suggests the possibility of a bent rotor. Observation of axial rotor movement should lead to a check of whether water end bearings are worn or loose in their housings.

Rotor bearing lubrication

If the bearings are grease-fed, grease should be coming out of the drain lines. If the bearings are grease-packed, they should be flushed and repacked every 6 months. If lubricated with circulating oil, bearing seals are checked for leaks, and the pump for level and oil temperature. The oil must be checked and changed at regular intervals, and the oil filter properly maintained.

Dust stops

Leakage and operating temperature are checked. To determine wear with Farrel yoke hydraulic (FYH) dust stops, the distance between the rotor endplate and gland ring backing is checked. (FH, F, FY and FYA dust stops are similar.) With self-sealing and adjusting (SSA) dust stops, the distance between the rotor endplate and the spring retainer is checked. (SS and SST dust stops are similar.)

Drop door and latch

The door should open quickly to a 135° angle then decelerate to the fully open position. The closing action should begin quickly and decel-erate over the last few inches. It should not slam shut. The actuator should be checked for leaks (particularly if there is a deviation from the door movements). The latch sliding stroke is checked by measuring the movement of the microswitch trip rod, which should be 2.5–2.625 in. The latch housing should be observed during mixing and when the latch engages. Movement indicates loose bolts or keys, or a cracked bed plate. The door and latch limit switches are checked. Once set, they should not require adjustment. Rollers and trippers should be clean to prevent faulty signals. If the switches fail to make contact during closing, the doortop and side edges should be checked for compound hangup. Safety hooks or bolts for the drop door are checked, as well as the presence of a legible safety tag.

The door is then opened and safety hooks or bolts engaged, and the hydraulic unit shut down with the accumulator drain open. Water piping and hoses are checked for leaks and wear. The latch should be posi-tioned so its front is flush with the front of the housing. If the contact plate is worn, it should be reversed or replaced (making sure the bolts are locked). There should be 0.010 in clearance between the door and the door support. Lock wires should be in place, as well as the door grounding cable. Thermocouple(s) should be secure with tips in place and wiring in good condition.

Hydraulic system

The accumulator drain and master shutoff valve are checked, as well as the presence of legible safety tags. The pressure gauge is then calibrated. Observing one mixing cycle, pressure drop should not be greater than 300 psi. The accumulator precharge is checked; it should be about 50 psi higher than half the system pressure. The filler breather unit should be in place. The oil level and condition is then checked. Contaminated oil is the most common cause of damage to hydraulic systems. Oil should be changed and the filter cleaned every 6 months. Replaceable cartridge filters should be changed as indicated by the build in pressure drop. Hydraulic oil should be kept in a contamination-free location. The preventive maintenance program should include hydraulic pump motor lubrication as per the manufacturer's instructions.

Grease system

Fresh grease should be in evidence around door shaft bearings, latch and bearing seals. If not, the grease supply in the pump reservoir should be checked. The high pressure burst disk is checked; if the disk has burst, the system is checked for blockage and repaired. The grease pump stroke adjustment is checked, as well as the needle valve (or restrictive orifice) that controls the flow of hydraulic fluid to the pump.

Dust stop lubrication

Pump motors and drives are checked for overheating, noise and vibration. The fluid levels in drive gearboxes are checked. The rate of flow of lubricant through the dust stops is checked against any recommendations. Too low a rate can cause overheating and premature wear to the dust stops. Too high a rate will leak lubricant into the batch. The oil supply is checked for contamination (which can damage the pumps and lead to dust stop failure).

Drive gears

Noise and vibration are observed while idling and under load. Oil level and gear condition are checked, as well as gearbox filters, pressure and temperature gauges. Uneven gear wear, excessive noise or vibration suggests an immediate check on component alignment.

Couplings

Couplings are checked for leaking seals. The level and condition of grease is checked, as well as proper installation of coupling guards.

7.1.2 Inspection of the Banbury platform

Ram and cylinder

The ram should move quickly during upstrokes and downstrokes. An integral air check should cushion the ram on its upstroke. On its downstroke a rubber or plastic ring in the bottom of the cylinder eliminates metal-to-metal contact. This is augmented by a restrictive valve in the discharge air line that activates near the conclusion of the downstroke.

The top and bottom cylinder air lines are checked with test gauges for pressure differential. Leaking packings can usually be identified by close listening. The maximum recommended actual pressure on the compound is 60 psi (the mixer manual gives the relationship between air pressure and actual pressure for a particular ram configuration). The effect of ram pressure on mixing is discussed in detail in section 1.6.1.

Floating weight

Regardless of design, the ram weight should 'float' on the piston rod. Lubrication of the point of attachment (weight rod cavity) is checked, as well as the condition of the dust cover.

Piston rod

The rod is observed during mixing. Excessive side-to-side movement indicates worn bushings, which make it impossible for the rod packings to hold air. This condition can lead to rod breakage. The piston rod is checked for evidence of scoring. Any packings and bronze bushings will wear rapidly.

Weight pin assembly

The pin assembly in the side plate that locks the weight in the up position (for maintenance or cleaning) is checked for condition.

Hopper door

The speed of operation of the hopper door is checked and, if necessary, adjustment made to the flow control on the air valve. Air cylinders are checked for leakage and proper lubrication. Side plate safety pins (which lock the door in the open position) are checked. Powder leakage at door edges can be corrected by adjusting seal strips (PTFE seals) or by cleaning (bronze seals). Underdoor leakage can be corrected by adjusting the seal assembly (as per the mixer manual).

Air line filter

The air line filter is checked for clean and dry condition.

Hopper operation

The recommended procedure is to close the hopper door before lowering the ram. This prevents materials being driven into the door by the action of the ram weight. Ram pressure should be relieved (ram floated) before opening the drop door. This prevents damage to the drop door and reduces hangup. Dropping of loose powders to the mixed batch can be reduced by leaving the ram in the down, but floating, position until the drop door is closed.

These inspection steps can be carried out quite rapidly by a trained technologist. Periodic inspection, a suitable preventive maintenance program and an appropriate lubrication schedule combine to maximize the useful life of the mixer, minimizing unscheduled downtime and the costs that accompany it.

7.2 MIXER MAINTENANCE AND LUBRICATION

- Points of lubrication are given in Fig. 7.1.
- Recommended commercially available lubricants are given in Fig. 7.2.

7.2.1 Each time the mixer is started

1. Check the lubricant grease pump reservoir is full.
2. Check the drive couplings are lubricated.
3. Check the lubricant level in the gearbox is correct.
4. Check the fluid level in the hydraulic unit is correct.
5. Check the dust stop lubrication reservoirs have an adequate supply.

7.2.2 Once per shift

1. Check the pressure gauges are operating.
2. Check the grease lubricating system is operating and the level is correct.
3. Check the dust stop lubricators are functioning and the levels are correct.

POINTS TO BE LUBRICATED	METHOD OF APPLICATION	LUBRICANT RECOMMENDED	FREQUENCY REPLENISH/CHANGE
MAIN BEARINGS GREASE PACKED (replenishing)	Hyd. Grease Pump	FG-2M	Each Shift
MAIN BEARING SEALS			
DISCHARGE DROP DOOR SHAFT & LATCH			
DUST STOPS A. Sealing Surfaces Only B. Flushing Cavities	Auto-pump	F4DS *Customer	As Required As Required
HOPPER DOOR SHAFT	Hyd. Grease Pump	FG-2M	Each Shift
PISTON ROD PACKINGS (Hopper Cover)			
Floating Weight	Manual Grease Gun	FG-2M	Daily
LOW SPEED Coupling & HIGH SPEED Coupling	See vendor data		
ROTARY/SWING JOINTS CARBON BEARINGS BALL BEARINGS	Grease Gun	FG-2M	8 Weeks
AIR SUPPLY LINE	Air Line Oiler	F2H or F2	Weekly / As Required
HYDRAULIC SYSTEM	Reservoir / Pump	F2H	Weekly / 6 Months
UNIDRIVE	See vendor data.		

*Oil must be compatible with stock. Flash point must be 300° F minimum. Viscosity not to exceed 3500 S.S.U. at 60° F.

Figure 7.1 Points of lubrication. (Courtesy the Farrel Corporation)

4. Check the drive lubrication is functioning and the levels are correct.
5. Check the hydraulic system fluid level is correct.
6. Check the air system is working properly.
7. Check the floating weight, drop door, door latch, hopper door and dust stops are functioning properly.

7.2.3 Once per day

1. Grease all points not lubricated automatically.
2. Check the dust stops for quantity of lubricant flow.
3. Check lubrication of the discharge door and latch, the hopper door, the weight rod and any seals.
4. Check for replacement of indicator lightbulbs.

FARREL® LUBRICATION CHART

FARREL® Lube Number	F-2H	F-2	F-6	F-11EPT	F-4DS	FG-2M
Viscosity CST.	61.2/74.8 @40°C	61.2/74.8 @40°C	288/352 @40°C	73/98 @100°C	414/505 @40°C	2 ①
Viscosity SUS (REF)	317/389 @100°	317/389 @100°F	1533/1881 @100°F	340/460 @ 210°F	140/160 @ 210°F	
Max. Operating Temp. °F	225 ②	225 ②	250 ②	225 ②	250	275
CHEVRON	AW HYD OIL 68	GST OIL 68	AW MACH. OIL 320	NL GEAL COMP.2200	AWTURBINE OIL 460	POLYUREA EP-2
EXXON	NUTO H68	TERESSTIC 68	TERESSTIC 320	SPARTAN EP 2200	TERESSTIC 320	LIDOK EP 2
TRIBOL	943 AW-68	943 AW-68	1100/320	MOLUB ALLOY 876 LIGHT		MOLUB ALLOY 860/150-2
MOBIL	DTE 26	DTE HVY MED	DTE AA	MOBIL TAC MM	VACTRA HH	MOBILITH AW 2
TEXACO	RANDO OIL HD 68	REGAL OIL R & O 68	REGAL OIL R & O 320	MEROPA 3200	REGAL OIL R & O 460	MULTIFAC EP 2
SHELL	TELLUS OIL 68	TURBOT OIL 68	MORLINA OIL 320	MALLEUS COMPOUND	MORLINA OIL 460	RETINAX LC GREASE 1

① CONSISTENCY - NLGI

② BASED ON 180° F MAX.SUMP TEMP. REFERS TO LUBRICANT ONLY,
FOR MAXIMUM OPERATING TEMPERATURE OF EQUIPMENT REFER
TO INSTRUCTION MANUAL.

Figure 7.2 Commercial lubricants. (Courtesy the Farrel Corporation)

7.2.4 Once per week

1. Check all filters and lubricant condition.
2. Grease the connection between the floating weight and the rod.
3. Check the air line filter and lubricator.
4. Check the hydraulic system; verify the pressures are correct.
5. Clean the area around the dust stops (do not disturb the dust stops if they are operating properly.

7.2.5 Once per month

1. Inspect the discharge door and latch; examine the door for possible misalignment.
2. Inspect the floating weight assembly; examine it for air leakage.

3. Inspect the cooling system strainers.
4. Examine all pipe and hose connections and their fittings.

7.2.6 Every six months

1. Drain and flush all oil reservoirs, including the hydraulic unit; clean all filters and refill the reservoirs with oil.
2. Examine the condition of drive gear teeth.
3. Flush the water system.
4. Inspect the mixing chamber and rotors.

7.3 ANTICIPATING REQUIRED SERVICE

Normal wear increases the clearance between the rotor tip and the wall, reducing the shear rate at a given rotor speed. Rounding of the rotor tip leads to the same characteristics. In their early stages these effects are often successfully combated by increasing the batch size. On the other hand, normal wear leads to a slight drop in the fully down ram position. This effect reduces optimum batch size. Worn dust stops permit leakage of compound and raw materials; this is accentuated by rotor wear and rounding. Variations in specific compounds with long mixing histories may be used as indicators. This complements but does not eliminate the need for regular inspection.

A number of the most common day-to-day problems are listed in the troubleshooting guide of Fig. 7.3

7.4 DUST STOP MAINTENANCE

FYH hydraulic dust stops are currently standard on Banbury mixers with volumes up to 270 liters (F-270); larger mixers use the FH type. SSA dust stops are available as alternatives.

The SSA dust stop is shown in Fig. 7.4. The floating and collar rings turn with the rotor; the replaceable wear ring is stationary (secured to the endplate). The spring mounting of the collar permits limited axial movement of the rotor; it is sealed at the floating ring by a PTFE or fluoroelastomer O-ring. Lubrication to the wear faces of the floating ring and the wear ring is fed through the wear ring. Proper flow rates of lubricating oil versus rotor speed are given in Fig. 7.5. And to prevent powder accumulation, process oil is flushed through the gap between the rotor and the endplate; plasticizer may be substituted for process oil in some nonblack rubber compounds. After flushing, the process oil usually contains some dissolved polymer, but sometimes it is collected and used as an ingredient in noncritical compounds. The flow of

TROUBLESHOOTING TABLE

The following table describes typical maintenance problems and their solutions:

PROBLEM	PROBABLE CAUSE	POSSIBLE SOLUTION
Ram does not come up.	1. Leaking packing around cylinder piston. 2. No air pressure. 3. Faulty controls. 4. Material sticking in hopper neck.	1. Repack 2. Correct condition. 3. Correct condition. 4. See next problem.
Material sticks in hopper neck.	1. Accumulation of material. 2. Oversize batch. 3. Improper loading of liquids.	1. Clean hopper neck. 2. Reduce batch size. 3. Use more care loading.
Material sticks to rotors or discharge door.	1. Machine temperature not correct. 2. Indicated batch temperature not correct. 3. Improper loading sequence. 4. Ram pressure too high.	1. Vary machine temperature within limits of material. 2. Check batch with pyrometer. Check thermocouple. 3. Vary sequence. 4. Reduce pressure.
Stock temperature rises too quickly.	1. Ram pressure too high. 2. Improper batch size. 3. Improper loading sequence. 4. Cooling system not working properly.	1. Reduce pressure. 2. Resize batch. 3. Vary sequence 4. If necessary, clean system.
Stock temperature does not rise as required.	1. Ram pressure too low. 2. Improper batch size 3. Improper loading sequence.	1. Increase pressure. 2. Resize batch. 3. Vary sequence.
Discharged batch is not satisfactorily mixed.	1. Improper batch size. 2. Improper loading sequence 3. Basic ingredients not of good quality.	1. Resize batch. 2. Vary sequence. 3. Verify with Q.C. or supplier.
Excessive leakage through dust stops.	1. Incorrect lubrication. 2. Incorrect adjustment for wear 3. Worn parts. 4. Poor face contact.	1. Correct flow rates and supply. 2. Check assembly drawing. 3. Replace. 4. Adjust contact.

Figure 7.3 Typical maintenance problems and suggested solutions. (Courtesy the Farrel Corporation)

lubricant and the flow of process oil should both be carefully monitored. Recommended flow rates of process oil are plotted in Fig. 7.6 for a typical rubber compound and in Fig. 7.7 for a typical plastic compound.

The SSA dust stop has been in use for more than 30 years in rubber and plastics compounding with outstanding success in sealing and noted ease of maintenance. The wear and floating rings need to be replaced as a pair and lapped together (as in a new installation). The wear ring is readily removed after disconnection of lubrication lines and removal of the clamping bolts. The floating ring is best removed by replacement of the pressure springs (Fig. 7.4) with screws and steel plugs to provide jack points. And heat is typically applied. After loosening from the rotor, it is common to cut through the rings with a torch to facilitate removal.

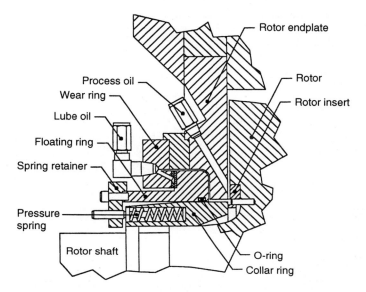

Figure 7.4 SSA dust stop. (Courtesy the Farrel Corporation)

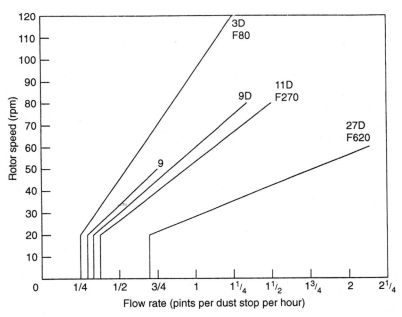

Figure 7.5 SSA dust stops: rotor speed versus flow of lubricating oil. (Courtesy the Farrel Corporation)

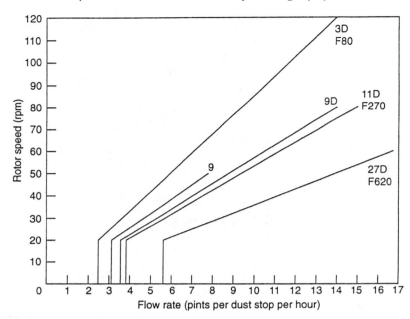

Figure 7.6 Rubber compounds with SSA dust stops: rotor speed versus flow of process oil. (Courtesy the Farrel Corporation)

Following removal of old dust rings, the endplate cavity, collar ring and all keyways must be thoroughly cleaned. The replacement procedure is described in detail in section 7.5. After dust stop replacement and mixer inspection, the new parts should be seated by running the machine without any load; an 8 h run-in period is common.

Diagrams of the FYH and FH hydraulic dust stops are given in Figs 7.8 and 7.9. These designs date from 1968 and were intended to provide easier maintenance and replacement. They have proven particularly long-lived in the high speed mixing of highly loaded rubber compounds. But with proper preventive maintenance, either SSA or hydraulic dust stops may be used in any Banbury, either for rubber or plastic compounds. Many plastics and some rubber compounders continue with SSA dust stops based on their particular experience.

In the FYH dust stop (Fig. 7.8), the wear ring and the clamp ring turn with the rotor. The wearing is an insert clamped into a recess at the end of the rotor by means of a collar ring. Sealing is against the stationary gland ring made of hard bronze alloy backed with steel. Axial rotor movement is accommodated by sealing with an O-ring. The entire assembly is held in place by an inverted Y-shaped yoke which operates under hydraulic pressure. Positioned at the top of the inverted Y-frame,

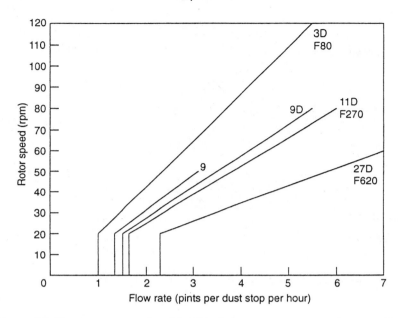

Figure 7.7 Plastic compounds with SSA dust stops: rotor speed versus flow of process oil. (Courtesy the Farrel Corporation)

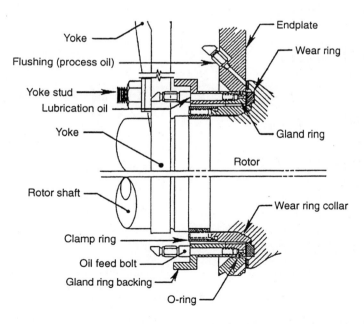

Figure 7.8 FYH dust stop. (Courtesy the Farrel Corporation)

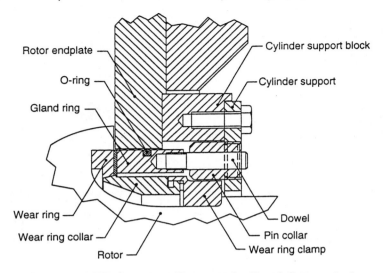

Figure 7.9 FH dust stop. (Courtesy the Farrel Corporation)

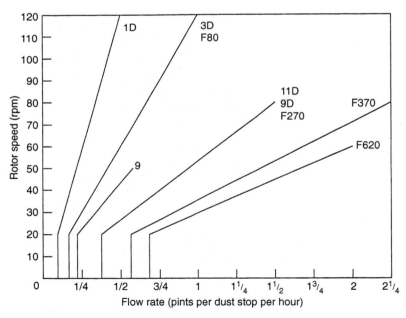

Figure 7.10 Hydraulic dust stops: rotor speed versus flow of lubricating oil. (Courtesy the Farrel Corporation)

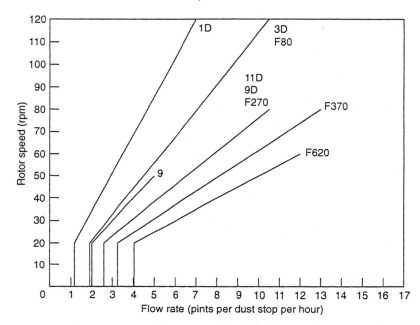

Figure 7.11 Rubber compounds with hydraulic dust stops: rotor speed versus flow of process oil. (Courtesy the Farrel Corporation)

the cylinder exerts its pressure using the mixer frame as its fixed point. The center of the yoke acts as a fulcrum for lever action by the ends of the Y-frame, and this action holds the dust stop in place.

In the case of the FH dust stop (for large mixers) the Y-yoke is replaced by an assembly of hydraulic cylinders bearing directly on the gland ring. This system is designed so that axial rotor movement is accommodated by shifting hydraulic pressure in correspondence, reducing vibration and wear.

In both cases the lubricant is fed through the gland ring. Recommended lubricant flow rates for hydraulic dust stops are plotted in Fig. 7.10. Process oil (or plasticizer) is introduced through the rotor endplates. Figure 7.11 plots recommended flow rates for typical rubber compounds and Fig. 7.12 for typical plastic compounds.

Replacement of hydraulic dust stops usually means no more than fitting a new bronze gland ring (as long as proper lubrication has been maintained). This may be done with little difficulty. After disconnecting hydraulic and lubrication lines, the fulcrum attachment is loosened and the yoke lifted out. After removing the securing bolts, the backing ring may be taken out. The bronze gland ring is normally cut with a torch to facilitate removal from the rotor. After cleaning the cavity, the wear

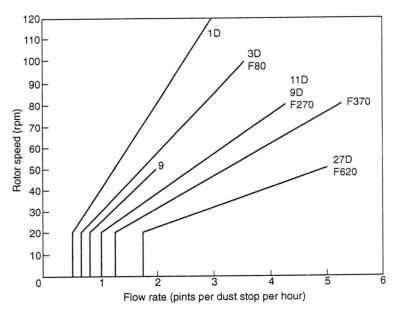

Figure 7.12 Plastic compounds with hydraulic dust stops: rotor speed versus flow of process oil. (Courtesy the Farrel Corporation)

ring is inspected (usually it is found in acceptable condition). If replacement is required, the clamp and collar rings must first be removed. Replacement components must be positioned correctly. Professional assistance may be advisable if more than the gland ring requires replacement, then the opportunity could be used for complete inspection. The normal replacement procedure is given in section 7.6.

Depending on the size of the Banbury, as many as 32 pumps may be involved in delivery of lubricating oil to each application point. These should be installed neatly, with care not to confuse lubricating and process oil lines.

It is recommended that the flow of process oil is continued between batches. If a delay of more than 2 min is expected, this should be shut off to prevent deposition into the mixing chamber. Such action can be initiated automatically, using a timing relay and solenoid actuated by the 'ram up' limit switch. If the ram remains up for longer than say 2 min, the timer trips and opens a bypass valve in the process oil pump system. Lowering the ram reverses the process and resets the timer. Other devices have been constructed to operate in a similar manner.

Modern Banbury mixers use distinct oil pumping systems for lubricant and process oil. These may be motor driven or hydraulically operated

(Hydrolube) systems. They are much more suitable for high speed, high pressure mixing than older systems that used a single pump with dual reservoirs; the older systems sometimes fail under high back pressure. Older systems may be converted easily. On the other hand, SSA and hydraulic dust stops cannot be interconverted without major rebuilding.

During steady mixing, SSA dust stops should normally run at 250–280 °F, as measured by probing the back of the wear ring, except that #9 mixers should instead show 200–240 °F. A measurement of over 300 °F is consistent with inadequate oil flow. Under the same conditions, hydraulic dust stops should run at 140–170 °F, except #9 mixers run at 110–130 °F. In this case the reading is from the back of the backing ring or the gland ring. Temperatures should not exceed 200 °F with proper oil flow. If the flow of process oil is held to a minimum, to avoid leakage into the compound, the flow of lubricating oil should be increased in compensation. This can be monitored by a check on the operating temperature.

Adherence to these simple principles can be relied upon to maximize the service life and productivity of any Banbury mixer.

7.5 SSA DUST STOPS

7.5.1 Assembly

1. Ensure that rotor collars, floating rings and keys, wear ring dust stop cavities in rotor endplates are completely free from rust, protective compound, dirt, etc. There should be no nicks, blemishes, etc., on any edges or surfaces.
2. Ensure that floating ring keys are securely staked into rings, square to faces; spring holes in rotor collars have no sharp edges or internal projections and springs have no sharp edges at either end.
3. Cut O-rings to length as specified in Table 7.1. This will prevent buckling.
4. Check that the floating ring is free to float, using a pry bar.

Table 7.1 O-ring sizes to prevent buckling (Courtesy the Farrel Corporation)

Size	O-ring length (in)	Spring deflection (in)
3A to 3D	27.75 + 0.66 − 0.06	9/16
9	36.63 + 0.00 − 0.06	9/32
11, 11D F-270	46.06 + 0.00 − 0.06	11/32
27D F-620	69.56 + 0.00 − 0.06	15/32

5. Remove one spring pressure screw on each dust stop and measure the depth from the outside of the spring retainer to the top of the spring or spring button. To this, add the desired spring deflection from the chart (which will give 22 psi on the seal face). Also measure the threaded length of the screw if used with spring only, or the overall length if spring buttons are used on top of the spring. Calculate the screw setting as follows:

Assume depth to be	5/8 in
Assume required deflection to be	11/32 in
Total	31/32 in
Length of screw measures	29/32 in

Setting in relation to face of spring retainer = length of screw − total of deflection and hole depth = 29/32 − 31/32 = − 1/16 = − 1/16 below the surface of spring retainer

7.5.2 Lapping

1. Disconnect all lube and process oil lines.
2. Remove wear rings, coat seal face with #360 carborundum compound and replace. No coarser grit than #280 should be used, and it is advisable to add a little oil with this grade.
3. Run for 1 h maximum, then remove top half of each ring and check for contact, which must be continuous round face and be at least two-thirds of total face width. If top halves are satisfactory, remove bottom halves and check contact. If top halves are not acceptable, recoat, replace and run again.

Repeat process till satisfactory contact is obtained.

4. Some rings may bed in sooner than others. In this case, thoroughly clean out all traces of grinding compound and rebuild dust stop with lube oil hooked up as detailed in item 1 of section 7.5.3.
5. Temperatures should be ⩾ 15 °F and not more than 50 °F above ambient. If the minimum is not obtained, the rings are probably not in contact. Check the floating ring action as per item 4 of section 7.5.1.
6. Before starting production in the machine, run dust stops with lube oil as specified in item 1 of section 7.5.3 for as long a period as possible but not less than 8 h, whether or not lapping has been carried out.

7.5.3 Running

1. Set all lube and process oil pumps to deliver at least the amount of oil indicated in Figs 7.5 to 7.7 for the maximum Banbury mixer rotor speed.
2. Ring temperatures should not exceed 300 °F during normal operation.

7.6 BANBURY MIXER: HYDRAULIC DUST STOPS

7.6.1 Assembly

With the hydraulic cylinders in place but with no hydraulic pressure, tighten the yoke stud nut sufficiently to deflect the yoke spring by 3/16 in. This will allow approximately 5/16 in of spring travel before the rod endguide bottoms in the endframe. This does not apply to the F-620 mixer, where cylinders are directly on the gland ring.

7.6.2 Run-in

Set the hydraulic pressure to obtain 15 psi on seal faces; set the lubrication oil flow to approximately one-half the level recommended in Figs 7.10 to 7.12. To expel air, crack bleed plugs at each cylinder and

Table 7.2 Hydraulic pressure to develop the required face pressure (Courtesy the Farrel Corporation)

Mixer size	Run-in or lapping pressure (psi)	Initial production pressure (psi)	Probable final pressure (psi)	Cylinder pressure/seal face pressure
F-620	210	900	560	14.3
F-370	210	800	450	14.0
F-270	145	700	400	9.5
11D	145	700	400	9.5
9D	160	700	400	10.25
9	115	600	350	10.22
F-160	–	–	–	–
F-80	145	550	350	10.0
3D	145	550	350	9.51
1D	130	550	350	8.75

on the machine-mounted piping. To achieve best results, this may have to be repeated several times, even after production has started in the mixer. Run with the mixer empty at the highest speed available for as long as possible, but not less than 8 h. Table 7.2 gives the hydraulic pressure to develop the required face pressure.

7.6.3 Lapping

Although not as essential as with SSA dust stops, lapping will ensure the best possible starting conditions for the dust stops and the Farrel Corporation recommends it be done in the best interests of the mixer user. Specific instructions are available.

7.6.4 Production

Set the hydraulic pressure as high as possible, depending on the door system operating pressure; set the lubrication and oil flow levels as shown in Figs. 7.10 to 7.12. After a few days of satisfactory operation, the hydraulic pressure may, at the customer's discretion be reduced by small increments, say 25 psi, at regular intervals, say 2 or 3 days to the lowest pressure consistent with satisfactory dust stop operation. It may be necessary also to adjust oil flows to obtain optimum conditions for a particular application. Provided lubrication is properly maintained at all times, high hydraulic pressure will not be detrimental to the dust stops. And note that process or lube oil flow at too high levels will induce leakage in hydraulic dust stops.

7.6.5 Flushing

With the mixer empty and running, release the hydraulic pressure. Adjust the flow of lubrication oil to its maximum. Run like this for a few minutes. When the hydraulic pressure and lubrication oil flow are returned to their normal settings, good sealing should be restored. But if there are hard deposits of stock on the seal faces, the mixer will probably need to be dismantled for cleaning of gland rings and rotor insert faces.

7.6.6 Other points

- Some older mixers are not equipped with process oil feeds; they may need a slightly higher hydraulic pressure and lubricating oil flow.
- Process oil cannot be tolerated in many plastic compounding applications. This may call for slightly higher hydraulic pressures and lubricating oil flows.

- Where process oil can be used without detriment, better results have been obtained using much lower flows than we recommend for rubber compounding.
- Consult Figs. 7.11 and 7.12 for suggested flow rates of process oil.
- This instruction is supplementary to those contained in Section 6 of the Instruction Manual for general startup procedures.

7.6.7 Hydraulic pressure ratio

See sections 7.6.2, 7.6.3 and 7.6.4.

7.6.8 O-ring assembly

Cut O-rings as per Table 7.3 for proper assembly.

Table 7.3 O-ring lengths for proper assembly (Courtesy the Farrel Corporation)

Mixer size	O-ring diameter (in)	O-ring diameter (mm)	O-ring length (in)	O ring length (mm)
1D	0.210	5.33	22.35/22.25	568.0/565.5
3, 3D	0.210	5.33	31.90/31.80	810.5/808.0
F-80	0.139	3.56	32.50/32.40	825.5/823.3
9D	0.210	5.33	46.60/46.50	1183.5/1181.0
11, 11D	0.210	5.33	49.60/49.50	1260/1257.5
F-270	0.139	3.53	48.88/48.78	1241.5/1239.0
F-370	0.210	5.33	59.95/59.85	1522.5/1520.0
F-620	0.210	5.33	73.60/73.50	1869.5/1867.0

8

Mixing procedures for specific compounds

Richard F. Grossman

8.1 INTRODUCTION

The following sections are devoted to a selection of actual compounds that have been mixed successfully in a #11D Banbury using the procedures given. This is not to say they are the optimum procedures for every internal mixer, or even the best for the machine that was used; improvements may well occur to the reader. This point will be further discussed later on. The specific compounds have been selected so as to illustrate or typify the features of mixing that must be kept in mind by the rubber technologist. Each section stresses the relationship between mixing and formulating, or perhaps revising a formulation. These elements of rubber technology can never be compartmentalized.

8.2 EPDM EXPANSION JOINT COVER

Table 8.1 describes a nonblack EPDM expansion joint cover which would be hand built from calendered (or extruded) strips or sheet, and molded over intermediate layers of fabric-reinforced SBR (or CR). With relatively soft nonblack EPDM compounds, conventional mix cycles are generally preferred to upside-down mixing with machines using two-wing rotors. Those with four-wing rotors could mix this compound upside down without incorporation problems. If the fillers were replaced with N500, 600 or 700 series carbon black, upside-down mixing would be the first choice. No one in the United States would

The Mixing of Rubber
Edited by Richard F. Grossman
Published in 1997 by Chapman & Hall, London. ISBN 0 412 80490 5.

Table 8.1 EPDM expansion joint cover

EPDM, amorphous, 60 Mooney	150 lb
Hard clay	125
Ppt. silica	50
Titanium dioxide	10
Zinc oxide	7.5
Low MW PE	7.5
Paraffin	7.5
Petrolatum	7.5
Polyol wax	2.5
Stearic acid	1.5
Paraffin oil	75
Sulfur	1.25
MBTS	1.25
TMTD	1.5
DPTT	1.5
TeDEDT, 85%	1.5

Coolant	Set at 120 °F, mixed at 25 rpm
0 min	Load rubber, 50 lb clay, small ingredients, except polyol wax
1 min	Load rest of clay and 1/2 oil
2.5 min	Load rest of oil, ppt. silica, polyol wax
150 °F	Load curatives
Dump	Dump at 225–235 °F after about 6 min
Mill	Front roll warm, back roll cool

mix this recipe in two passes unless there was a specific need for un-accelerated compound (for storage, or for use with several cure systems).

Several features of this compound are worth noting. The high levels of lubricants used (1 phr each of petrolatum, paraffin, and low MW polyethylene), presumably to aid processing in building the expansion joint cover, would be difficult to incorporate if added late in the cycle. If a lower level were used (in some other application), it would be added later, to maximize its effects in further processing.

Polyol wax is added to surface treat the silica filler. Such a treatment should always be added with the filler it is meant to coat, not with other ingredients (with which it may complex instead).

A combination of time and temperature parameters is typically used to decide when the mix is complete with compounds like the material in Table 8.1. Very little dispersive work is involved here, and the level of power draw is mainly an indication of proper batch size.

8.3 EXPANSION JOINT INTERMEDIATE LAYER

Table 8.2 considers a typical intermediate layer for an expansion joint. Like the material in Table 8.1, this SBR compound may be routinely mixed in a single pass unless the unaccelerated compound is required by other considerations. The procedure illustrates the use of SBR black masterbatch as an aid in mixing. The first stage is a sandwich mix, converting the SBR 1778 clear polymer to compound by incorporation of filler and oil, broadening its solubility parameter, facilitating the blend with SBR black masterbatch. The second addition of ingredients makes use of the remaining SBR 1805 to speed incorporation of curatives. In this case the resultant cycle is much more rapid than if a conventional procedure had been followed (blending the elastomers then adding

Table 8.2 Expansion joint intermediate layer

SBR 1805 (2×90 lb bales)	180 lb
SBR 1778 (1.5 bales)	112.5
Whiting, dry ground	75
Hard clay	50
Zinc oxide	8
Coumarone–indene resin	7
Antioxidant	3
Stearic acid	2
Naphthenic oil	35
Sulfur	3.25
CBTS	3
TMTM	0.5

Coolant	Set at 120–125 °F, mixed at 20–22 rpm
0 min	Load SBR 1778, filler and oil, small ingredients Load first bale of SBR 1805
1.5 min	Load curatives; load last bale of 1805
Dump	Dump at 220–225 °F after about 4–4.5 min and at an established range of total power consumption
Mill	Front roll neutral, back roll cool

increments of filler and oil). As long as the overall oil content is modest, this type of modified sandwich mix can be carried out with mixers of older design (sliding door, spray cooling) providing that the equipment is well maintained (clearances are within specifications). This is in contrast to batches with high oil and low polymer content, as with many EPDM compounds, which are convenient in upside-down or sandwich mixes only in internal mixers of modern design. With four-wing rotors, on the other hand, all ingredients might be added at time zero, in a sandwich load. Typical levels of overall power consumption by a specific mixer should be determined by observation for compounds of this type.

8.4 TRAFFIC COUNTER TREADLE COVER

Table 8.3 deals with the mixing of a compound for encapsulating traffic counter treadles (such as used in highway tollbooths). In this

Table 8.3 Traffic counter treadle cover

Natural rubber, 80 Mooney	130 lb
Polyisoprene	150
SBR 1846	45
Ppt. silica	75
Zinc oxide	30
Resorcinol–formaldehyde resin	7
Zinc laurate	3
Antioxidant	3
Antiozonant	2
Metal deactivator	1
Magnesium oxide, high activity	4
Pine tar	6
Sulfur	6
OBTS	3
Hexamine resin	3

Coolant	Set at about 100 °F, mixed at 20 rpm
0 min	Load rubber
1 min	Load zinc oxide, small ingredients, 25 lb silica
2 min	Load pine tar, rest of silica; load cures
Dump	Dump at 220–230 °F after about 5 min; overall power consumption highly dependent on lot of NR
Mill	Front roll warm, back roll neutral

combination of natural rubber, synthetic polyisoprene and SBR black masterbatch, the relatively small quantity of the last ingredient argues against a sandwich mix (as might be used with 100 lb more SBR, 100 lb less NR or IR). The blend of natural and synthetic polyisoprene provides a broad molecular weight (MW) distribution that improves certain processing operations, such as preform extrusion from a hot feed extruder with low L/D ratio (length-to-diameter ratio).

This compound must provide high adhesion to the steel workings of the traffic counter mechanism, which prompts the use of silica filler, high zinc oxide level and pine tar as a tackifier, as well as resorcinol–formaldehyde resin. Stearic acid is then commonly replaced with zinc laurate as a cure activator, decreasing the release agent function. Fine particle silica is used conveniently to help incorporation, particularly of pine tar. Compounds such as this naturally have good adhesion to steel surfaces in general (such as mill rolls). For efficient processing, it is necessary to mix consistently, dump at a consistent temperature then to remove the batch promptly from the mill. A batch of this compound could easily pose problems in removal if it were dumped with unmixed zinc oxide, if it were substantially overmixed or if it were allowed to remain on the mill during a coffee break. Compounds based on natural rubber will tend to exhibit lot-to-lot variation in overall power consumption. In critical applications this is overcome by extensive crossblending. In cases where the natural rubber is broken down before use, as is typical with smoked sheet, the breakdown procedure can be modified to accommodate incoming lot-to-lot variation in viscosity. The product can then be blended before mixing of the compound. The variation is smaller with premasticated and controlled grades and, anyway, many of them are crossblended from different lots. Establishment of useful limits for overall power consumption depends on the extent to which the incoming variability is controlled. This is not to say that the power consumed (i.e. the work done) is not of the greatest importance in mixing compounds of this type, merely that its control is not necessarily simple.

8.5 SBR/IR BELT COVER

Where different types of SBR black masterbatch are used in the same compound, the more filler and oil extended grade is usually added last. This is illustrated in Table 8.4, a common SBR/polyisoprene belt cover formulation. Here a relatively small quantity of highly reinforcing carbon black must be well dispersed. In the procedure adopted, the batch begins with a sandwich load: clear rubber, carbon black and other ingredients that are (relatively) difficult to disperse, then the SBR 1600 type masterbatch on top.

Table 8.4 SBR/IR belt cover

SBR 1606	180 lb
SBR 1848	135
cis-Polyisoprene	75
N347 HAF	30
Zinc oxide	7.5
Hydrocarbon resin	7.5
Antioxidant	3
Antisuncheck wax	3
Stearic acid	3
Sulfonated oil	5
Sulfur	5.5
OBTS	3.5
TMTM	0.5

Coolant	Set at 120 °F, mixed at 20 rpm
0 min	Load IR, black, resin, zinc oxide; load 1606
1 min.	Load 1848
2 min	Load oil, waxes
160–165 °F	Load curatives
3–3.5 min	
Dump	Dump at 220–230 °F after 4.5–5 min and meeting established overall power consumption standards
Mill	neutral front roll, cool back roll

After 1 min the softer SBR 1800 series masterbatch is added. Experience with these batches has shown that reversing the order of the 1600 and 1800 additions lowers the degree of black dispersion and resultant physical properties. Even with this complex procedure, the mix cycle is quite rapid and the output high. If instead the three elastomers are preblended followed by the addition of carbon black, almost all of the black will be associated with the SBR phase. In the present procedure, the added HAF is distributed at least partially in the polyisoprene. This is not to say that the overall balance of properties will necessarily be more suitable (although this is true for the belt cover application), but that it will be different depending on the path. With consistent monitoring of incoming raw materials and adherence to the mix procedure, this type of compound should yield an overall mix cycle highly consistent in batch time, dump temperature and overall power consumption.

8.6 EPDM LOW VOLTAGE ELECTRICAL CONNECTOR

Mineral-filled EPDM compounds are often mixed conventionally. If a two-pass mix cycle is to be used, the first pass can often be mixed upside down, with increased output, even with two-wing rotors. The compound in Table 8.5 is mixed in two passes because (1) it enables convenient screening of the first pass and (2) the mineral rubber requires mixing temperatures of >300 °F for good dispersion. (Mineral rubber is a plasticizer having good electrical properties. It is used here, along with hydrocarbon resin, in partial replacement of paraffinic process oil, to maximize adhesion to other layers of the connector or to the electrical inserts.) The cure system, although relatively safe, could not survive temperatures reached in the first pass.

The employment of a brief holding period after addition of filler and oil, before addition of the polymer, tends to speed incorporation and

Table 8.5 EPDM low voltage electrical connector

EPDM, amorphous, high ENB	150 lb
Calcined clay	225
Mineral rubber	25
Hydrocarbon resin	15
N550 FEF	5
Zinc oxide	7
Zinc stearate	0.5
Paraffin wax	5
Paraffinic oil	45
Coolant	Set at 180–200 °F, mixed at 30–35 rpm
0 min	Load filler, oil, small ingredients Hold 30 s; load rubber
Dump	Dump at 320–325 °F after 7–8 min; batch over 300 °F for 1–2 min
	Second pass
Sulfur	1.5 lb
MBT	0.75
MOTS #1	1.5
Coolant	Set at 100–120 °F, mixed at 20 rpm
0 min	Load 1/2 first pass (cool); load curatives Load rest of first pass
Dump	Dump at 220–225 °F after 2–2.5 min

smooth breakdown of the bales. In the first pass, standard batch time, dump temperature and overall power consumption will be highly consistent with these ingredients. In the second pass, as is common, cure ingredients are sandwiched between additions of first pass. The first pass is best allowed to cool by standing overnight; testing is completed before it is crossblended.

8.7 PEROXIDE-CURED BLACK-FILLED EPDM COMPOUNDS

Two quite similar compounds are given in Table 8.6. Compound A is a jacket for a molded electrical connector; compound B is a gasket for a steam pressure gauge. Compound A has been mixed without difficulty in a single pass, assuming a modern internal mixer with good cooling and which is not excessively worn. Certainly technologists equipped only with older mixers, perhaps with spray cooling, would choose to mix compound A in two passes, adding the peroxide curative in a second pass.

Two factors tend to reverse the situation with compound B: use of FEF instead of SRF as the major filler, and the inclusion of methacrylate monomer coagent. With sulfur cures instead, most rubber technologists (at least in the United States) would mix this compound in one pass. The combination of peroxide plus coagent, taken with the higher shear

Table 8.6 Peroxide cured, black-filled EPDM compounds

	Compound A	*Compound B*
EPDM, amorphous, low Mooney	225 lb	225 lb
N774 SRF	150	50
N550 FEF	–	100
Zinc oxide	11	11
Paraffinic oil	25	35
Zinc stearate	1	1
Methacrylate coagent, 75% dispersion	–	6
Dicumyl peroxide, 40% dispersion	18	20

Single-pass	Upside-down load at 20 rpm
	Coolant at about 100 °F
	Dump at 215–220 °F
Two-pass	Load masterbatch upside down at 30 rpm
	Coolant at 150–180 °F
	Dump at 265–275 °F

involved in mixing FEF, suggests mixing in two passes. (With close supervision, a single-pass mix is not out of the question.)

Coagents (additives with reactive double bonds) may be added to the first pass of nonpolar polymers if the dump temperature is kept below about 280 °F. This is no longer the case if the polymer eliminates reactive molecules such as hydrogen chloride during processing. If the coagent is added to the second pass, it should be kept separate from the peroxide and the two mixed in consecutively. This also applies to blends of peroxides. It should not be presumed that cross-linking ingredients will remain inert when blended with each other under high shear in an internal mixer. This is particularly the case when one or more is liquid and potentially flammable. Combustion is very unlikely to occur during mixing of compound 8 or similar compounds, but where both coagent and peroxide are added to the same pass, it is standard practice to incorporate the coagent before adding the peroxide.

8.8 EPDM CONCRETE PIPE GASKET

A superficially similar compound is given in Table 8.7. Although this compound uses FEF black and methacrylate coagent, it would usually be mixed in a single pass. Two factors predominate: the high level of process oil holds down the temperature during mixing, and the use of bis(*t*-butylperoxy)diisoproylbenzene in place of dicumyl peroxide increases process safety. With this bis-peroxide, EPDM compounds may generally be mixed to 225–235 °F without danger of premature vulcanization. These limits depend on the reactivity of the base polymer; with

Table 8.7 EPDM concrete pipe gasket

EPDM, amorphous, high Mooney	175 lb
N550 FEF	150
Zinc oxide	7
Methacrylate coagent, 75%	5
Paraffinic oil	110
Methacrylate coagent, 75%	5
bis(*t*-butylperoxy)diisopropylbenzene	18

Coolant	Set at 100–120 °F, mixed at 25 rpm
0 min	Load black and oil, small items
	Hold 30 s; load rubber
150 °F ~ 3 min	Load peroxide
Dump	Dump at 220–225 °F after 5–6 min

peroxide-cured SBR or NBR, mix temperatures above 220 °F should be avoided with all currently available peroxide initiators. Peroxides more active than dicumyl (e.g. having lower decomposition temperatures) are invariably mixed into EPDM in the second pass of a two-pass mix. Such initiators, usually ketone bis-peroxides, restrict the mix temperatures to 165–185 °F (a good rule is not to approach within 100 °F of the desired cure temperature).

8.9 SBR INSULATION

A common mixing situation arises with the compound in Table 8.8; it is desired to mix rapidly until a certain temperature is reached, in this case, about 300 °F, then to slow the mix cycle to give several minutes of mixing at 300–325 °F. Here this is to ensure incorporation of high melting mineral rubber and styrene resin; in other cases, alternative goals may suggest the same strategy. A minor quantity of a liquid or low melting (but not a readily ignitable) solid ingredient can often be added to cut the shear rate and maintain the batch within the desired temperature range; stearic acid will serve this purpose. A soft antioxidant or other antidegradant should not be used for this purpose; it should be added early in the mix cycle to protect the polymer. Only as a last resort should the temperature be

Table 8.8 SBR insulation

SBR 1503 (2 bales)	150 lb
SBR 1018 (1/2 bale)	37.5
Calcined clay	250
Mineral rubber	75
High styrene resin	35
Zinc oxide	25
N774 SRF	5
Antioxidant	6
Litharge, 90%	3
Stearic acid	3

Coolant	Set at 180–200 °F, mix at 35 rpm
0 min	Load rubber, clay, all ingredients except mineral rubber and stearic acid
1 min	Load mineral rubber
300 °F ~ 4 min	Load stearic acid
Dump	Dump at 325 °F after about 6–7 min

controlled by rotor speed variation within the mix cycle. Even if they are automated to eliminate operator variability, these cycles are undesirable because of the extra wear and tear on drive components. Nevertheless, there are certain situations where it becomes necessary to adjust the rotor speed during the batch, often to control the manner in which the mixer discharges the batch to downstream equipment.

Variations of the compound in Table 8.8 were used over many years for underground transmission of low voltage power in telephone communications, but they have been superseded in the United States by EPDM insulations.

8.10 INJECTION-MOLDED NBR GASKET

NBR compounds for mechanical goods tend to have optimum property sets when mixed in two passes, with the first pass mixed at 275–300 °F. Since this appears to be the case with both black and clay-filled stocks, the cause is unlikely to be the interaction of heat with the filler. But whatever the cause high quality NBR compounds are usually two-pass mixed. It is important to achieve the best possible sulfur and zinc oxide dispersion in NBR; these are typically added with the polymer. Surface-treated sulfur and coated grades of zinc oxide are used. Active plasticizers often interfere with sulfur and zinc oxide incorporation, so plasticizer and stearic acid are added after these ingredients have been mixed in.

In NBR (and polar polymers generally) thermal black is more difficult to disperse than 'soft' furnace blacks; thermal black is therefore added early. The more readily dispersed SRF is used to help incorporate plasticizer and stearic acid. But suppose that no such obliging ingredient is present, i.e. suppose the only filler in the compound is a low level of MT black (Table 8.9). Then it is necessary to mix in the sulfur, zinc oxide and antioxidant, followed by the black. The batch has to be mixed with multiple increments of plasticizer and stearic acid in a long, inconvenient cycle that reduces the most modern internal mixer to the efficiency of a prewar model. Such is all too often the result of injudicious compounding.

Accelerator loadings can be substantial for high speed injection molding, particularly at low sulfur levels. Nonetheless, a modern internal mixer with controlled cooling can accommodate a compound such as the NBR of Table 8.9 without reduction in batch size for the second pass. Older mixers, such as with spray cooling, would use a reduced second pass (as would all equipment with very active curatives). In designing a reduced second pass, consideration should be given to the length of the average campaign with the compound in question. For example, the compound in Table 8.9 theoretically produces 469 lb of product per batch

Table 8.9 Injection-molded NBR gasket

NBR, medium ACN, medium Mooney	225 lb
N762 SRF	150
N990 MT	50
Zinc oxide	7
Stearic acid	1
Sulfur, treated	0.5
Antioxidant	3.5
Dioctyl phthalate (DOP)	32

Coolant	Set at 130–140° F, mixed at 25–30 rpm
0 min	Load rubber, MT, 50 lb SRF, sulfur, zinc oxide, antioxidant
1 min	Load rest of SRF, plasticizer, stearic acid
Dump	Dump at 280 °F after 5–6 min; set values of power input

Second pass

TMTD	2.7 lb
TETD	2.7
TBBS	3.7

Coolant	Set at 80–90 °F, mixed at 20 rpm
0 min	Load ½ first pass; load curatives; load rest of first pass
Dump	Dump at 175–185 °F after about 2 min

of first pass. Experience has indicated some loss of volatiles, dust and scraps of compound. The factory average loss may be perhaps 2%, suggesting a net of 460 lb per batch. In order to supply manufacturing with 10 000 lb of finished compound, the mixing room would probably schedule 22 batches of first-pass mix. (An even number would be scheduled if the 225 lb of polymer per batch represented 4½ bales each of 50 lb, as is typical with NBR. With 75 lb bales, either an odd or even number of batches might be scheduled. With 50 kg bales, about 110 lb, the batch would have been based instead on 220 lb, or 100 kg, of polymer. All of this assumes a certain level of rational behavior.) The 22 batches should produce between 10 100 and 10 150 lb of first pass. The exact quantity will have been noted by the quality control department when it decided how to crossblend the first pass.

Probably 25 batches of second pass would be scheduled, calling for 405 lb of crossblended first pass, 2.3 lb each of TMTD and TETD, and

3.2 lb of TBBS. After actual weighing of the first pass, possibly 24 batches of the above weights might be generated; an unforeseen shortage might cause batch 25 to be readjusted to 387 lb of first pass plus 2.1 lb of TMTD and 2.1 lb of TETD along with 3.5 lb of TBBS. In such a way, no dribs and drabs of raw materials or first pass remain (to be discarded or worked away inappropriately into the next compound).

Addition of sulfur to the first pass is typical of NBR compounding, although unusual with most other elastomers, and reflects the difficulty in incorporating sulfur into NBR uniformly. It is less safe, and is rarely carried out when NBR is blended with other polymers such as SBR or BR. Fortunately, these blends tend to have broader solubility parameters than pure NBR, so they require less effort to incorporate the sulfur.

When sulfur is added in the second pass, it should be in the form of a dispersion.

8.11 CR/SBR BLEND

The formulation in Table 8.10 is a common polychloroprene/SBR blend used for calendered sheeting. Products where the CR content is about half the polymer content are often called **commercial neoprene**, i.e. they conform to specifications requiring the polymer content to be at least 50% polychloroprene. These blends are quite satisfactory for many applications.

With a solubility parameter of about $17 \, \text{MPa}^{1/2}$, SBR is less polar than CR (solubility parameter of about 19). However, the solubility parameter of SBR is broad. SBR/CR blends tend to have high compatibility in all proportions. Therefore the two elastomers may be added together (as they are in Table 8.10). Generally no plastic contamination other than special low melting ingredient bags may be used in CR batches because of the low dump temperatures used and the sparing solubility of polyolefins in CR. In molded CR applications the presence of plastic contamination appears as blisters, often elongated and generally containing a pinpoint of fluxed plastic. This appearance is characteristic and can be distinguished by the experienced technologist from blisters caused by trapped air or ingredient outgassing during cure. Thus, if the compound in Table 8.10 were to be used in molding, the mix card would require removal of all plastic, such as the SBR bale wrapper. In these cases, even low melting ingredient bags should be investigated before being used routinely; plastic-contaminated CR is almost impossible to recover.

In the case of autoclave cure (hose, belting, calendered sheeting) or steam CV (wire and cable), blistering from plastic contamination in CR is usually not a problem (the applied pressure is not sufficient to trap and freeze in the discontinuities). Hence no such precaution is included

Table 8.10 CR/SBR blend

CR, unmodified, high Mooney	75 lb
SBR 1502	75
Whiting	200
N774 SRF	75
Factice	15
Low MW PE	8
Zinc oxide	6
Magnesium oxide, 85%	3
Petrolatum	3
Antioxidant	2
Stearic acid	1.5
Aromatic oil	46
Sulfur	3
TMTM	1.5
Coolant	Set at 100 °F, mixed at 20 rpm
0 min	Load all rubber, whiting, all small items
1 min	Load SRF and oil
150 °F	Load sulfur and TMTM
~ 4 min	
Dump	Dump at 210–215 °F after 6–7 min

with the compound of Table 8.10. Quality control testing of CR compounds for molded applications should check for plastic contamination by visual observation of molded buttons or slabs (often the effect is not seen in cured rheometer samples).

The same lack of solubility makes it possible to use low molecular weight polyethylene as a release agent for the compound. SBR/CR blends are typically more tacky than pure CR equivalents; roll adhesion can be a problem with low durometer compounds. Hard clay is therefore best avoided, but Factice helps.

Sulfur/TMTM cures, often amine activated, are commonly used with SBR/CR blends and lead to smooth simultaneous vulcanization. With sulfur cures, zinc oxide can usually be added early in the batch. If added towards the end, a zinc oxide dispersion is usually employed. Recall that MBTS and TMTD are not scorch improvers in sulfur-cured CR, and certainly not in SBR/CR blends). ETU and other thiourea systems are not easy to control in SBR/CR systems unless only a minor proportion of SBR is present. If more than a minor fraction of SBR is involved, the properties provided by thiourea cure systems, superior to sulfur in pure CR, are no better in SBR/CR blends.

8.12 LOW DUROMETER CR/SBR BLEND

The compound of Table 8.11 is a low durometer (40A) composition for molded goods. In this case, although SBR and CR are compatible in terms of polarity, the contrast between the relatively hard CR chips and the soft 1800 series SBR suggests the use of a sandwich mix. SBR 1848 (100 parts SBR, 82.5 HAF, 62.5 aromatic oil) is used as a tool to aid incorporation of other ingredients, particularly the high level of Factice. The procedure of Table 8.11 is not satisfactory for soft printing rolls or blankets (where the utmost in dispersion is needed); a CR/Factice

Table 8.11 Low durometer CR/SBR blend

CR, unmodified, medium Mooney (2 bags)	110 lb
SBR 1848 (2.5 × 90 lb bales)	225
Factice	100
Platy talc	50
Zinc oxide	10
Magnesium oxide, 85%	4.5
Antioxidant	3.5
Antisuncheck wax	2
Petrolatum	2
Stearic acid	1
Aromatic oil	6
Sulfur	2.5
TMTM	2.5

No plastic into batch

Coolant	Set at 80–100 °F, mixed at 20 rpm
0 min	Load CR, filler, zinc oxide, antioxidant, magnesium oxide, ½ factice
	Load 1.5 bales of SBR 1848
1 min	Load oil, waxes, rest of factice, rest of SBR
180 ° F	Load sulfur and TMTM
~ 3.5 min	
Dump	Dump at 220 °F after 6–7 min

masterbatch, using premium grades of Factice, is often mixed instead. For ordinary mechanical goods, the single-pass mix used here is usually adequate.

Mill release is rarely a problem at Factice levels as high as this. The combination of wax, stearic acid and petrolatum serves as a lubricant to promote further processing and to carry antioxidant to the surface during service.

The rubber technologist should keep up to date with the range of premixed masterbatches supplied by polymer producers. Although intended for large consumers such as tire manufacturers, these master-batches should be considered as starting points for compounds in general. For example, mixing the compound of Table 8.11 from pure SBR polymer and oil would be very tedious and probably involve mixing temperatures high enough to force second-pass addition of cures.

8.13 CR SPECIFICATION COMPOUNDS

Compounds A and B of Table 8.12 have been employed in fabrication of parts that conform to government specifications. Compound A is

Table 8.12 CR specification compounds

	Compound A	Compound B
CR, unmodified, medium Mooney	275 lb	187.5 lb
N990 MT	50	150
N774 SRF	75	150
Zinc oxide	14	–
Magnesium oxide	11	7.5
Antioxidant	7	4
Stearic acid	2	1
Butyl oleate	48	40
ETU, 95%	3	–
TMTD	2	–
No plastic into batch		
Coolant temperature	90 °F	110 °F
Rotor speed	20 rpm	20 rpm
0 min	Load CR, MT, small items	
1.5 min	Load SRF and plasticizer	
160 °F	Load curatives, if single pass	
Dump	Dump at 220 °F after 6–7 min	

intended for extrusion and is about 55A in hardness. Compound B has been extruded and lathe cut into gaskets; it is about 80A hardness. CR compounds above 70A hardness are commonly mixed in two passes when thiourea cure systems are used.

With a conventional internal mixer equipped with two-wing rotors, the rubber technologist would almost certainly mix CR masterbatch as indicated by the procedure of compound B. This is a conventional mix cycle that uses a semireinforcing filler, SRF black, to aid incorporation of plasticizer. With an internal mixer equipped with four-wing rotors, the shear rate is such that compound B, or a similar first pass, could readily be mixed upside down with improved output. This would not be the case, however, if a fine particle black (e.g. HAF) were used. Fortunately, the combination of thermal and coarse furnace blacks tends to provide an excellent balance of properties with CR. If, by way of unpleasant example, the recipe for compound A were to consist of fine particle carbon black and ester plasticizer, a two-pass mix would be needed.

Compounds based entirely on CR handle on the mill much like natural rubber analogs, generally preferring a warm front and cool back roll, and quickly incorporating traces of unmixed ingredients that may drop with the batch.

8.14 CHARGE-DISSIPATING CR TUBING

Most rubber articles that dissipate static charge (so as to prevent spark discharges in sensitive environments) do so by being intrinsically semi-conductive rather than through the use of antistatic agents at their surface. Many antistatic agents, widely used in thermoplastics, are simply too soluble to function in semipolar rubber compounds. General practice is to add sufficient conductive carbon black to achieve a surface resistivity of 10^4 to 10^8 Ω cm adequate to drain surface charge even at low relative humidity. A second filler is often used to moderate the effect of the conductive black, as in the compound of Table 8.13. The blended compound is commonly found to process more readily. Besides the second filler, processing of sulfur-modified CR is often assisted by a peptizing agent, piperidinium pentamethylene dithiocarbamate (PPDC). This reagent should be added with the rubber, ahead of other additives; in particular, it should not be combined with amine antioxidants. On the other hand, it is necessary to achieve uniform distribution of the conductive black and breakup of any coarse aggregates. Therefore the black is added shortly afterwards, before a great deal of viscosity reduction can occur. The need to delay addition of other ingredients produces a slow mix cycle, but this is convenient because attempting to mix semi-conductive compounds at high output typically leads to escape of fine particle black and its distribution throughout the factory.

Table 8.13 CR tubing, charge dissipating

CR, sulfur modified (4.5 × 55 lb bags)	247.5 lb
N293 CF	100
Whiting, dry ground	50
Zinc oxide	13
Magnesium oxide	16
Low MW PE	5
Octylated diphenylamine	4
Stearic acid	2
PPDC (peptizer)	1
Aromatic oil	42
Coolant	Set at 80–90 °F, mixed at 20 rpm
0 min	Load CR and PPDC
1 min	Load black, zinc and magnesium oxides Close hopper; drop ram slowly
3 min	Load oil, whiting, rest of small items
Dump	Dump at 210–215 °F after about 8 min

The peptizing agent PPDC is a strong accelerator for unsaturated elastomers and should be used with great caution in blends of SBR and sulfur-modified CR. It has essentially no effect on unmodified CR. The compound of Table 8.13 uses zinc and magnesium oxides for cross-linking, without organic accelerators. This is common with semi-conductive sulfur-modified CR, as is the high level of magnesium oxide, both of them helping to increase scorch resistance.

This type of compound more commonly uses a coarser black such as FEF or SRF (in conjunction with fine particle black) instead of whiting or other mineral filler. Addition of minor amounts of the coarse black (e.g. 10–15%) often improves the speed of incorporation of the finer particle black. This effect, known to rubber technologists for many years, may relate to improved efficiency of mechanical surface wetting, or it may be involved with dewetting of small molecules (usually water) from the black surface. Anyhow, coarse black is very useful when it is inappropriate to choose a high structure black that is easily incorporated.

With polar elastomers, particularly NBR and epichlorohydrin elastomers, there may be no need to use special conductive blacks. Before proceeding with this type of compound, the technologist should investigate whether it might instead be adequate to employ about 40 phr N347 (or similar black). Should this be the case, it would generate much higher rates of output.

8.15 NONBLACK CR FOR INJECTION MOLDING

CR compounds for injection molding, particularly of relatively small parts, typically use sulfur-modified polychloroprenes with thiourea type cure systems to achieve fast cure rates. The compound of Table 8.14 can easily output small parts in 30–40s cycles at 375–400 °F in a modern injection press. A black compound would most likely use SRF as filler, whereas nonblack compounds commonly use mixtures of clay and whiting. To develop the best possible tear strength in a resilient nonblack CR, a combination of hard clay and precipitated silica is often used. The

Table 8.14 CR, nonblack, for injection molding

CR, sulfur-modified	220 lb
Cis-4 1203 BR	15
Hard clay	150
Ppt. silica	25
Polyethylene glycol, add to silica	0.5
Zinc oxide	11
Magnesium oxide	9
Low MW PE	6
Antisuncheck wax	5
Antioxidant, nonstaining	4
Stearic acid	1
Petrolatum	1
Naphthenic oil, low volatiles	38
ETU, 95%	2
TMTD	2
No plastic into batch	(BR must be stripped)

Coolant	Set at 80 °F, mixed at 20 rpm
0 min	Load rubber, small items, 50 lb clay
1 min	Load oil, rest of clay
150–160 °F	Load ETU and TMTD
~ 3 min	
Dump	Dump at 210–215 °F after about 5 min at specific total power consumption

addition of BR is also helpful in maximizing hot tear strength, an important factor in the rapid part removal needed in injection molding (particularly when complex parts must be demolded from cores to which they have greater than zero adhesion).

Hard clay imparts tackiness to CR compounds, particularly sulfur-modified CR. Therefore, to maintain internal strength, this type of compound should not be overmixed; power consumption should be monitored. In practice, mixing instructions might call for completion of the batch at over 200 °F but below 220 °F, after a specific total power draw.

With sulfur-modified CR, TMTD functions not only to retard scorch but also to accelerate the rate of cure, once cross-linking has started. This is appropriate for high speed molding of small parts, where the mold often starts to open before 90% cure is reached, depending on thermal momentum to drive the cure to completion.

8.16 HARD RUBBER INDUSTRIAL WHEEL

The compound given in Table 8.15 (70–80D hardness, when fully cured) could serve either as the core beneath a softer tread stock, or as a one-

Table 8.15 Hard rubber industrial wheel

SBR 1502	75 lb
SBR 1009/1018	25
Barytes	100
Whiting, dry ground	100
Hard clay	50
Hard rubber dust	100
Naphthenic process oil	45
Sulfur	50
Hydrocarbon resin	10
Stearamide (process aid)	5
Amine activator	1

Coolant	Set at 90–100° F, mixed at 20 rpm
Equipment	All equipment grounded for static charge
0 min	Load fillers, resin, rubber, then load sulfur
1 min	Load rubber dust and oil
160 °F	Load process aid and activator
Dump	Dump at 215–220 °F after about 6–7 min (225 °F max)
Sheet-off mill	Cool front, warm back roll

piece caster, depending on the application. SBR 1009 or 1018 is included for control of swell and shrinkage during extrusion of preforms for molding. With hard rubber compounds, these must be essentially on size if cracking and backrinding during molding is to be avoided, as there is a substantial exotherm during cure.

Compounds with levels of sulfur above 10–15 phr are usually mixed (carefully) in a single pass, rather than in two passes; this is because it is difficult to incorporate these sulfur levels into completely mixed compound with uniform distribution. There is little to choose in terms of scorch safety before or after adding activators or accelerators to a high sulfur batch of hard rubber. If the batch begins to cure during mixing (260–280 °F or above), it will almost certainly ignite. No one should undertake such mixing without a second (backup) thermocouple, thorough grounding against static discharge and the presence of appropriate fire extinguishers. With the proper precautions it is easy to mix simple hard rubber compounds such as that of Table 8.15. Modified late oil addition with ground hard rubber dust (section 2.4.1) is common with internal mixers having two-wing rotors. With the four-wing configuration, an upside-down order of addition is feasible (assuming normal cooling efficiency); the batch size would probably be reduced to lower further heat buildup.

8.17 HIGH DUROMETER NBR MASTERBATCH

With typical acceleration, including a cross-linking agent for the phenolic resin, the compound of Table 8.16 will develop a cured 90A hardness, yet enable relatively easy processing. This is a result of compounding with phenolic resin, which lends processability before the cure, and hardness afterwards (generally at the expense of resistance to compression set). It is analogous in this regard to compounding SBR with high styrene masterbatch. If the phenolic resin used already contains cross-linking activation, the dump temperatures should be held at 250 °F. This is somewhat inadequate for the compound of Table 8.16 because of the need to provide optimum distribution of sulfur, zinc oxide, phenolic resin and HAF black; a nonactivated resin is therefore chosen. When, in the second pass, accelerators (e.g. MBTS, TMTD) and a phenolic cross-linking agent (e.g. hexa plus activator) are added, they are best added separately rather than premixed. With mill addition of curatives, hexa dispersion is almost always used.

The mixing of compounds of this type can be greatly simplified by use of NBR/carbon black particulate masterbatches. For example, 150 lb of 100/75 NBR/SRF masterbatch, 115 lb of 100/75 NBR/HAF masterbatch, and 50 lb SRF black would provide ratios quite similar to the compound of Table 8.16. These could be preblended with sulfur, zinc

Table 8.16 High durometer NBR masterbatch

NBR, high ACN, medium Mooney	150 lb
Liquid NBR	10
N326 HAF-LS	50
N762 SRF-LS	125
Phenol–formaldehyde resin	38
Zinc oxide	5
Low MW PE	5
Stearic acid	1
Antioxidant	3
Treated sulfur	1.5
Dibutyl phthalate	20

Coolant	Set at 140 F, mixed at 25–28 rpm
0 min	Load rubber, N326 and sulfur
1 min	Load zinc oxide and phenolic resin
2.5 min	Load N762, oil, rest of ingredients
Dump	Dump at 265 °F after about 6–7 min
Mill	Cool front roll, warm back roll

oxide and phenolic resin then mixed rapidly with late oil addition of liquids and soft ingredients.

8.18 NBR/PVC CABLE JACKET

If the desired application were cord jacket, the compound of Table 8.17 would normally be mixed in a single pass with curatives (by 'normally' I mean that the mixing equipment on hand is not worn out or decrepit). However, brief inspection indicates that the polymer content is too high and the plasticizer too expensive for such an application. For use instead as heavy-duty cable jacket, the optimum properties of the composition depend on the mixing step. Here this involves good MT black dispersion, so the MT black is added early. Intermediate blacks (FEF, SRF, GPF) pose less of a dispersion problem and can be used to assist plasticizer incorporation. It may be safely assumed that the more effective is a plasticizer in promoting low temperature flexibility, the more it will tend to 'break' the batch (i.e. incorporate slowly). The FEF/plasticizer mixture, on the other hand, will incorporate relatively rapidly. If FEF

Table 8.17 NBR/PVC cable jacket

NBR/PVC 70/30, medium Mooney	250 lb
N990 MT	100
N550 FEF	75
Hydrocarbon resin	10
Zinc oxide	7.5
Antioxidant	5
Antisuncheck wax	3
Stearic acid	1
Treated sulfur	0.25
Dioctyl sebacate	45
Coolant	Set at 150–165 °F, mixed at 30 rpm
0 min	Load rubber, MT and small ingredients
1.5 min	Load plasticizer and FEF
Dump	Dump at 285–290 °F after about 5–6 min with established level of work input

were replaced with a fine particle black (e.g. HAF), this procedure could not be used; it would be necessary to resort to late 'oil' addition of the plasticizer.

NBR/PVC blends cannot reach optimum properties if any unfluxed PVC persists. The best quality compounds are therefore two-pass mixed with a high dump temperature in the first pass. And incoming NBR/PVC should be checked for unfluxed PVC by addition of several percent carbon black to a sample on a lab mill. Clear spots that refuse to incorporate black indicate a phase very rich in PVC; such lots are best rejected. As a last resort, they may be remixed above the melting point of the particular PVC used in the blend, often requiring mixer temperatures of 320–350 °F. If this is attempted, additional PVC stabilizer of the same type used by the supplier should be added (generally organo-tin or calcium/zinc because they are sanctioned by the Food and Drug Administration). The potential problem of unfluxed PVC is most severe with PVC grades that contribute the best physical properties to the blend, i.e. those with standard rather than low molecular weight. Despite various claims, there is no conclusive evidence that NBR/PVC made by latex blending is generally any better in this regard than blends prepared from the polymers in an internal mixer. Whichever method is adopted, it is advisable to mix critical compounds in two passes, the first with a hot dump.

8.19 NBR/PVC/SBR BLEND

Although SBR and PVC are themselves incompatible, NBR/PVC blends with SBR have been used for many years. The broad solubility parameter of SBR permits development of a compatible blend, provided that process oil is restricted to aromatic grades (or ester plasticizers). Such blends have been run in almost all proportions. When both are major constituents and when SBR black masterbatch can be used, the procedure given in Table 8.18 is rapid and consistent. It is a further example of a premixed ingredient, SBR black masterbatch, being used to promote the incorporation and distribution of the remaining ingredients.

8.20 BUTYL MASTERBATCH

As with most amorphous, high molecular weight polymers having high free volume (i.e. not overly resilient), butyl rubber is best mixed with a batch size that is 10–20% greater than would be calculated from considerations of compound specific gravity and viscosity. The compound of Table 8.19 uses a batch size for a #11 Banbury that is about 10% larger than would be chosen for the same recipe using SBR. With natural rubber or neoprene it is common to have a batch that is 10–15% smaller. The butyl batch is very convenient, particularly for the nonbulk black user, calling for even bales of rubber and bags of carbon black. This is

Table 8.18 NBR/PVC/SBR blend

SBR 1848	180 lb
NBR/PVC 70/30, medium Mooney	100
Hard clay	100
Zinc oxide	8
Antioxidant	2.5
Stearic acid	1
Treated sulfur	3
Aromatic process oil	40
MBTS	3
TMTD	1.5

Coolant	Set at 100–120 °F, mixed at 20 rpm
0 min	Load NBR/PVC, sulfur, zinc oxide, antioxidant Load clay and oil; load first bale of SBR
1.5 min	Load accelerators; load second bale of SBR
Dump	Dump at 220–225 °F after about 4 min

a significant consideration with butyl, as the large batch size makes for a rapid mix cycle that may be difficult to keep up with during long runs.

Butyl rubber masterbatch is mixed most efficiently with mixer temperatures in the range of 150–180 °F. Controlled temperature coolant (tempered water) is a great advantage. The mixer should not be heated by previously mixing natural rubber; traces of highly unsaturated polymers (NR, SBR, NBR) will absorb the curatives selectively and are said to **poison** butyl. (This is not true of neoprene; blends of butyl and neoprene may be used with cure systems appropriate for both, keeping in mind that zinc oxide must be delayed until the second pass. It is also irrelevant in connection with natural rubber/butyl blends used in the uncured state as adhesives.) The run should be preceded by a butyl cleanout, reserved only for such use: 3 bales of butyl, 200 lb of SRF or GPF black, and a 10 lb block of paraffin are appropriate ingredients for such a batch in a #11 mixer.

If the masterbatch uses carbon black of intermediate fineness, or fairly dense semireinforcing mineral filler (e.g. platy talc, hard clay, wollastonite), all ingredients may be added at the start of the batch. The conventional order of ingredient addition is preferred with two-wing rotor mixers; four-wing machines can mix butyl upside down without difficulty (despite the common opinion that butyl must always be mixed conventionally). Any 'hard' blacks, as in the compound of Table 8.19 need to be incorporated before addition of oil or stearic acid.

Use of mineral fillers and low polymer content (below 40%) can contribute to strong mill roll adhesion with a number of butyl compounds. This invariably becomes worse with protracted milling and as the batch cools. With butyl compounds it is best to band the batch

Table 8.19 Butyl masterbatch

Butyl, 3.5% unsaturation	225 lb
N550 FEF	100
N326 HAF-LS	100
Zinc oxide	11
Stearic acid	4
Paraffinic process oil	27
Coolant	Set at 150–160 °F, mixed at 35 rpm
0 min	Load rubber, HAF and zinc oxide
1 min	Load FEF, oil and stearic acid
Mill	Cool front roll, warm back roll

immediately and to start the strip off the mill on the first revolution of the batch into the nip. High roll speed is also helpful, and the mill operator must usually be an expert. As in so many cases, circular mill strip knives, mounted low on the front roll, are a great advantage, as is a variable speed sheet-off mill. Running any great quantity of butyl without such refinements will tend to generate feelings of frustration in the most good-natured mixing department.

8.21 BUTYL MASTERBATCH, HEAT INTERACTED

Unsaturated polymers can react with fine particle furnace blacks in the range 250–350 °F, causing a gel structure to develop and the viscosity to increase. With highly unsaturated elastomers, particularly natural rubber, this occurs readily above 250 °F and is rapid over 320 °F. Natural rubber masterbatch with a 200 or 300 series black can easily be permanently gelled if the mixing temperature rises too high; it may even ignite if the batch runs away. This reaction can be very useful if controlled; it may be inhibited by amine antioxidants and by stearic acid. Thus, contrary to normal procedure, stearic acid and waxy antioxidant may be added to natural rubber before incorporating HAF or ISAF in order to prevent or moderate gel formation.

For butyl the level of unsaturation is low enough to permit good control over the reaction. The lowest unsaturation grades are difficult to heat react, but medium or high unsaturation butyl interacts readily with fine particle black. Stearic acid and antioxidants should then be reserved for the second pass (Table 8.20). As the heat reaction proceeds best without interruption, the usual procedure is to add all ingredients, oil included, at the start of the batch, which is then run up over 300 °F as rapidly as possible. Interaction is controlled by time and temperature

Table 8.20 Butyl masterbatch, heat interacted

Butyl, 2.7% unsaturation	300 lb
N330 HAF	125
Zinc oxide	15
Low MW PE	12
Paraffinic process oil	22

Coolant	Set at 165 °F, mixed at 30 rpm
0 min	Load all ingredients
Hold	Hold over 320 °F for predetermined time, usually 1–1.5 min
Dump	Dump at 340 °F (350 °F max)

above 320 °F. Power draw will increase in the range 300–320 °F as gel formation builds; later it will level off or drop. At this point the batch is dumped.

Heat interaction is accelerated by use of about 0.1–0.25 phr of *p*-dinitrosobenzene, a quinoid curing agent, added with the polymer. With this additive, even low unsaturation grades of butyl may be heat treated with fine particle carbon black. Its use has decreased with increasing perception of nitroso compounds as undesirable additives.

Halobutyl polymers may also be heat interacted using the same procedures, with temperatures reduced by about 20 °F at all stages. Metal oxides should not be included in such masterbatches.

8.22 CHLOROBUTYL/NR BLEND

The compound of Table 8.21 has been used as a heat-resistant belt cover; it develops excellent adhesion to friction or skim stocks based either on natural rubber or neoprene. Even with blends with unsaturated elastomers as high as 50 wt%, relatively large batch sizes are typical, at least with compounds that are relatively easy to mix. If some of the ingredients were known to cause dispersion problems (or if the mixing equipment were marginal), the choice of batch size might be made 5–10% lower. If the Mooney viscosity is not disparate, halobutyl and natural rubber (or SBR) may be combined, either in a conventional mix cycle, or for four-wing rotor machines, in an upside-down mix. If instead the natural rubber had a Mooney viscosity of 100–120, it would typically be added first in a sandwich mix. Use of magnesium oxide as a stabilizer in halobutyl blends with NR mixed to high temperatures is common; blends with SBR would commonly be mixed satisfactorily at 250–265 °F.

In the second pass, the most consistent cure rates are obtained by first mixing zinc oxide (or maybe litharge) into the compound, followed by the curatives. As with all halogenated elastomers, good zinc oxide dispersion is important in developing optimum properties. But rapid incorporation is vital with mill addition of cures (in compounds designed with adhesion as a characteristic). Therefore, quick-incorporating paste dispersions of the ingredients are frequently used.

8.23 CSM CORD JACKET

The rubber technologist must routinely be able to mix an increasing number of specialty elastomers, and an interesting example is chlorosulfonated polyethylene (Hypalon). As with many members of this group, a relatively low molecular weight leads to a predominantly elastomeric character over a narrower range of temperature and composition

Table 8.21 Chlorobutyl/NR blend

Chlorobutyl, medium Mooney	100 lb
Premasticated NR, 60–80 Mooney	100
Hard clay	125
N990 MT	50
Ppt. silica	50
Antioxidant	2
Magnesium oxide, high activity	1
Stearic acid	1
Naphthenic process oil	30

Coolant	Set at 150 °F, mixed at 30 rpm
0 min	Load all rubber, clay, MT and magnesium oxide
1 min	Load oil, silica, other ingredients
Dump	Dump at about 280 °F after 4–5 min and characteristic work input

Second pass

First pass, crossblended	450 lb
Zinc oxide	10
Sulfur	4
MBTS	3
TMTD	0.2

Coolant	Set at 100–120 °F, mixed at 20 rpm
0 min	Load ½ first pass, zinc oxide
1 min	Load curatives, rest of first pass
Dump	Dump at 190–200 °F; allow 30 s cooling

than would be the case with say SBR or EPDM. This can be a disadvantage in some situations, such as when elastomeric character is needed for clean handling on a sheet-off mill but the limits of rubbery behavior have been exceeded. On the other hand, there are corresponding advantages: the potential for a relatively large batch, and the opportunity to blend grades having very different viscosities. The combination of two grades having a marked difference in viscosity would prompt a sandwich mix with SBR or EPDM, but CSM requires no special consideration.

Compounds such as the nonblack cord jacket given in Table 8.22 are universally mixed in a single pass. A conventional mix cycle is usually

Table 8.22 CSM cord jacket

CSM, medium Mooney	110 lb
CSM, high Mooney	110
Treated clay	150
Whiting, water washed	50
Red iron oxide	15
Magnesium oxide, high activity	11
Low MW PE	5
Pentaerythritol, 200 mesh	5
Stearic acid	2
Aromatic process oil	20
Diisodecyl phthalate	30
Sulfur	2
TMTD	4
Coolant	Set at 100–120 °F, mixed at 20 rpm
0 min	Load rubber, 100 lb clay, process oil
1 min	Load DIDP, rest of filler, small ingredients
170–180 °F	Load curatives
Dump	Dump at 195–200 °F to neutral mill

employed with a two-wing rotor configuration. An upside-down mix is entirely satisfactory with four-wing rotors, but the batch temperature must be closely watched (slightly smaller batches may be needed). Two considerations are important with CSM compounds of this type. Coolant temperatures must be set so that no condensation occurs in the mixer. Water mixed into a CSM compound will drastically reduce shelf life and process safety. Similarly, no water leakage from air cylinders into the mixer can be tolerated. (This can occur with a combination of worn seals on the ram piston and inattention to water removal from compressed air sources.)

The second consideration involves reasonable procedures for delivery of two greatly different plasticizers to a given batch. The ideal system is dual injection with provision for preheating high viscosity oils, but this is not as common as desirable. A single system with a mixing tank is also a reasonable approach, provided it is possible to achieve accurate delivery and homogeneous combination of the ingredients. The blend will incorporate more readily than adding one after the other; this is because the mixture generally has a broader solubility parameter. The procedure that leads to the worst consistency is attempting to run the heavy oil and the low viscosity plasticizer alternately through the same injection system.

When CSM compounds similar to that of Table 8.22, but involving greater potential return to the fabricator, are mixed in two passes, the first pass is usually run upside down, regardless of rotor configuration. This is true of most elastomers that are supplied in small, well-partitioned pieces, which therefore resist massing together. In other words, the compound self-adheres more readily than the polymer.

8.24 NONBLACK MILLABLE URETHANE

Millable urethanes are further examples of polymers that exhibit elastomeric behavior over a narrower range than general-purpose rubbers. But unlike chlorosulfonated polyethylene, urethanes are condensation polymers not addition polymers. When condensation polymers are mixed, particularly in the presence of vulcanization ingredients, care must be taken not to reverse the condensation process (i.e. to depolymerize the rubber). This is always a possibility with condensation polymers.

The compound of Table 8.23 serves admirably as an insert for a citrus fruit juicing machine because of the toughness and abrasion resistance

Table 8.23 Nonblack millable urethane

Millable urethane, medium Mooney	300 lbs
Ppt. silica	75
Titanium dioxide	50
MBTS	12
MBT	6
MBTS/zinc chloride complex	3
Zinc stearate	1.5
Oleamide (process aid)	1.5
Sulfur	4.5

Coolant	Set at 100–120 °F, mixed at 25 rpm
0 min	Load rubber, titanium dioxide
1 min	Load silica, load small ingredients
Dump	Dump at >200 °F (240 °F max) when power draw levels off and begins to decrease
Mill	Warm front roll, cool back roll

of polyurethanes. It can be safely mixed in one pass, or the sulfur omitted to extend shelf life. In either case it is necessary to watch carefully for signs of reversion, i.e. a decrease in power draw on completion of ingredient incorporation and distribution. At this point, usually in the range 220–240 °F, the batch must be dumped, otherwise continued mixing will convert it into low molecular weight soup. The ease of reversion is a function of polymer molecular weight; therefore, the viscosity of incoming lots should be inspected and the mix procedure adjusted if necessary. Black-filled compounds are somewhat less sensitive. Reversion resistance is improved by using CSM as the source of chlorine instead of a zinc chloride complex, but at the expense of the cure rate. Use of cadmium stearate in place of zinc also improves the reversion resistance, but is inappropriate for articles intended for food contact.

With peroxide instead of sulfur cure, urethane compounds are almost always mixed in two passes, the second pass for addition of a peroxide dispersion, typically mixed to 180–210 °F. In this case, i.e. without MBT, MBTS and zinc chloride, if the initial Mooney viscosity is normal, the first pass may be mixed routinely at 240–250 °F. The same procedure may be used for blends of millable urethane and NBR, which are quite compatible at ACN levels above 35%. Such blends are most often mixed in two passes, with either peroxide or sulfur added in the second pass (the sulfur/MBT/MBTS/zinc chloride system covulcanizes the blend). In this case, the first pass may be mixed to 280–290 °F, if needs be.

8.25 ECO MOLDING COMPOUNDS

Epichlorohydrin polymers and copolymers are more difficult to revert than polyurethanes, but rapidly lose their elastomeric character above 200 °F, particularly if plasticized. Despite recent advances in cross-linking technology, many parts molded to existing specifications depend upon the lead oxide/ethylene thiourea (ETU) cure. Both elements of the cure system are toxic, and both must have exceedingly uniform distribution in the compound to produce optimum properties. In many cases the relevant specifications are built around the optimum property set; consequently, both the lead component and the ETU must be incorporated while the polymer is in its range of elastomeric behavior. These ingredients cannot be incorporated effectively with low shear response.

The compound of Table 8.24 has been used as an oil and fuel resistant vibration damper in outdoor applications. The batch is large for the specific gravity, as is typical with marginally elastomeric polymers. In this case, with black reinforcement, dispersion of the lead salt curatives is relatively simple. They may be safely added as dustless powder dispersions or prills, prepackaged to the proper batch weight in low melting point bags. The power draw early in the batch is such that

Table 8.24 ECO molding compound

ECO copolymer, medium Mooney	300 lb
N550 FEF	150
Dibasic lead phthalate, 90%	15
Dibasic lead phosphite, 90%	20
Nickel dibutyldithiocarbamate	3
Zinc stearate	3
Stearamide (release agent)	3

Coolant	Set at 150 °F, mixed at 25–28 rpm
0 min	Load rubber, lead salts, nickel stabilizer
1 min	Load FEF
3 min	Load zinc stearate, stearamide
Dump	Dump at 280–285 °F and established power input

Second pass

First pass, crossblended	full batch
ETU, 90%	5 lb

Coolant	Set at 100 °F, mixed at 20 rpm
0 min	Load ½ first pass; load ETU; load rest of first pass
1.5–2 min	Dump when power draw begins to drop (200 °F max)

dispersions, instead, in EPM or EVA are also readily incorporated. Similarly, a second pass may be used for adding ETU as a prill, dustless powder or elastomeric dispersion. The ETU is added to thoroughly cooled first pass either on a mill or in an internal mixer.

Problems arise when a compound such as the example of Table 8.24 includes 10–20 phr of plasticizer. In such cases a strong power draw is found only for the first minute or two of the first-pass mix. This is sufficient to incorporate the lead additives homogeneously, particularly if added as 90% dispersions in the plasticizer of choice. The problem is in subsequent incorporation of ETU dispersions in a plasticized compound that provides hardly any shear. Addition of both ETU and lead salt to the first pass is difficult to carry out without inducing precure. Commonly the user adds a soft dispersion of ETU to the equally soft compound in a second pass. To the eye this provides dispersion, but final properties rarely reach optimum levels. This is particularly the case when ETU is dispersed in a binder incompatible with ECO, such as EPM. This approach works with firm compounds that provide mixing shear, but fails when solubility (or the lack thereof) is the only driving force towards incorporation. Then the best solution is to run two first-pass batches, one containing ETU (added early, while power draw is high), the other containing the lead salt or salts, again added early (both as dispersions). The second pass combines the proper proportions of both to form the finished compound. This approach, often called the **half-batch technique**, has been used for many years with intractable cure systems of all kinds. It involves no greater overall input of mixing time than most alternatives, but creates endless possibilities for human error. On occasion the procedure is invaluable. A common application in years past was the mixing of half batches of EPDM for extrusion and hot air tunnel cure (after recombination); they used outrageously fast cure systems because the vulcanization capacity was inadequate to meet the desired line speeds. Unfortunately, the use of compounding and mixing technology to compensate for poor choice of further processing equipment has been very common.

In connection with ECO molding compounds, note that amide release agents and process aids are typically synergistic with metallic stearates, tending to form one-to-one complexes.

8.26 POLYACRYLATE SHAFT SEAL

Many elastomeric compounds such as that of Table 8.25 – compounds designed to seal against rotating shafts or used in other applications where frictional abrasion or spalling is anticipated – attempt to lower

Table 8.25 Polyacrylate shaft seal

Acrylate elastomer	225 lb
N550 FEF	175
Graphite powder	25
Stearic acid	8
Antioxidant	5
Release agent blend	8
Detergent type curative	15

Coolant	Set at 120 °F, mixed at 30 rpm
0 min	Load rubber, FEF, antioxidant
1 min	Load rest, except graphite
2–3 min	Load graphite as power draw turns level
Dump	Dump at 265–285 °F when power draw declines

the dynamic coefficient of friction with surface lubricants. These include molybdenum sulfide, PTFE, and in the present case, graphite powder. More often than not, addition of such a lubricant in more than trivial quantities makes it very difficult to incorporate further ingredients. In the compound of Table 8.25, the development of optimum cured properties depends on homogeneous distribution of the carbon black reinforcement and the curatives. Graphite distribution is not quite as critical, so graphite is added after the other ingredients in the first pass, with the FEF dispersed before adding any soft materials.

A problem then arises over addition of sulfur, the remaining curative. Addition as a dispersion on a two-roll mill is often possible; in other cases the low surface energy contributed by graphite leads to poor milling and inadequate distribution of sulfur. Then it is better to use the half-batch technique and combine two first-pass mixes, one containing the detergent accelerator, the other containing sulfur. In one case, where optimum distribution of lubricant was also needed, the processor combined three first-pass mixes, containing curative, sulfur and graphite, respectively.

8.27 XLPE INSULATION

Low density polyethylene (LDPE) exhibits something of an elastomeric character in a relatively narrow range of temperature above its crystalline melting point; precisely how narrow depends on the molecular weight of the LDPE, the type and extent of branching, and the molecular weight distribution. Most current cross-linkable polyethylene (XLPE) compounds for wire and cable are actually based on EVA copolymers of fairly low vinyl acetate content. These process more rapidly than analogs based on LDPE homopolymer and are also more easily mixed.

As indicated in Table 8.26, a single-pass mix is most often used. This requires close temperature control; without the ability to maintain consistently the actual batch temperature in a 10–15 °F range, it is best not to attempt this type of mixing. The most common practice is to mix the resin alone, almost to the flux point, then to add all powders. An exception is with semiconducting black, where the resin is heated above the flux temperature to reduce airborne effluent of the fine particle black. But here one is usually less concerned about maximizing output than when using more ordinary XLPE grades.

At the flux point, carbon black, treated clay or other filler, and antioxidants are readily incorporated. As soon as this stage of mixing is essentially complete, the appropriate peroxide (or coagent for electron beam cure) is added. Traces of unmixed powders assist in its incorporation. When mixed routinely, liquid or molten peroxide is injected directly into the batch, using special-purpose equipment available from

Table 8.26 XLPE insulation

EVA, 6–8% VA	300 lb
N990 MT, nonpelleted	81
Phenolic antioxidant	1
Dicumyl peroxide (pure basis)	7.5
Coolant	Set at 140 °F, mixed at 25 rpm
0 min	Load EVA resin
175–180 °F	When power draw levels off as resin starts to flux, load black and antioxidant
210–220 °F	Load peroxide
235–240 °F	Dump to warm mill; batch temperature should be 250–260 °F max

peroxide suppliers. (The ordinary injection system should not be used.) But when these compounds are mixed on general-purpose lines, the peroxide (or coagent) is best added as a powder dispersion.

With XLPE compounds the flow of lubricant through the dust seals is held to a minimum. The procedure of Table 8.8 leads to little outward filler pressure; the batch size is generally modest (in the unfluxed state, the typical pellet form of LDPE leads to low apparent bulk density). A light naphthenic oil may be used for lubricating dust seals; for colors an ester plasticizer. With coagents in place of dicumyl peroxide, or with peroxides having higher decomposition temperatures, the batches may be mixed to 260–275 °F. Nonetheless, compounds of this type should always be checked in process for signs of incipient premature cross-linking (precure). Premature cross-linking, along with processability, may be monitored by diversion of a small sidestream of the pelletized mixed compound continuously to a laboratory extruder equipped with a tape or Garvey die.

When the polymer is available in powder form, instead of pellets, compounds such as that of Table 8.26 may be mixed by preblending all powders in a ribbon blender, followed by single addition of all ingredients to an internal mixer, as would be done with PVC or CPE. The batch size would then be normal rather than small.

8.28 FKM MOLDING COMPOUND

Fluoroelastomer compounds such as the example of Table 8.27 are typically mixed upside down when the polymer is supplied in chips or

Table 8.27 FKM molding compound

FKM polymer, cure-incorporated pellets	500 lb
N990 MT, nonpelleted	150
Magnesium oxide, high activity	15
Calcium hydroxide	30
Internal lubricant	5

Coolant	Set at 80–100 °F, mixed at 25 rpm
0 min	Load black, oxides, polymer on top
165 °F	Load internal lubricant
~ 1.5–2 min	
Dump	Dump at 230–240 °F after 3–4 min

pellets, but conventionally when the polymer is in slab form. As with CSM, partitioned pellets mass more easily with the upside-down load. It is recommended that occasional users mix only cure-incorporated FKM grades, where curatives have been previously added by suppliers and consistent cure activity checked. The distribution of the curative (bisphenol AF; phosphonium salt catalyst) masterbatches must be exceedingly uniform for development of optimum properties. When used with FKM polymers not containing curatives, or to modify cure-incorporated FKM, these masterbatches should be added with the polymer. Blends of cure-incorporated grades should be limited to those from the same supplier unless it is certain that identical curatives are present; blends of different phosphonium salt accelerators can have strange and unexpected effects on cure rate and scorch safety. With addition of the same curatives to a blend of noncure-containing FKM grades, no such restriction need apply.

Distribution of magnesium oxide and calcium hydroxide is similarly important. Only thoroughly dry powders or dispersions in FKM should be used (such dispersions are best made with noncure-incorporated grades, or with liquid FKM, as binders). As with XLPE, nonoil-pelleted carbon black grades are more readily incorporated.

FKM compounds are highly intolerant of hydrocarbon process oils; therefore, dust seal lubrication should be held to the minimum flow needed, and ester plasticizer (DOP) substituted for naphthenic oil. Hydrocarbon rubber cleanout batches are totally unsuitable except as a prelude to hand cleaning of the internal mixer. After hand cleaning of the mixer and the areas above the ram that accumulate residue, a batch of black-filled polar elastomer (assuming black FKM batches are to be run), such as ECO or high ACN NBR, may be run and saved for future use. This batch should contain no additives, merely elastomer and black; in particular, all amines and sulfur compounds must be avoided. A dummy FKM batch may be run, but this is an expensive alternative.

It is generally more practical to mix FKM compounds in a smaller internal mixer. Batch sizes are about 65–70 lb in a #1; 240–270 lb in a #3 mixer. Sometimes it is possible to dedicate such a machine to FKM mixing, by far the most preferable route. Alternatively, mixed compounds may be purchased from special-purpose custom compounders, often with assurance of conforming (after proper vulcanization and postcure) to various specifications. Many processors who mix general-purpose compounds prefer to employ others to custom-mix FKM or similarly expensive elastomers.

8.29 SILICONE SPARK PLUG BOOT COVER

Filler-extended silicone compounds such as that of Table 8.28 are easily mixed in the internal mixer if several important precautions are followed. As with FKM, silicone compounds have low tolerance of hydrocarbons, either in oil or polymer form. Nonetheless, amorphous EPDM is silicone compatible and may be used for cleanouts if the fillers used are also suitable for silicones. This is limited to various forms of silicas, silicates and metal oxides. Clay and calcium carbonate should be avoided. Therefore, a cleanout batch ahead of a run of red silicone compound, e.g. spark plug covers, could use amorphous EPDM, silica (or calcium or magnesium silicate) and red iron oxide. A small amount of a low color, aromatic oil could be included; aromatic oil is also preferred for use in dust seal lubrication.

Table 8.28 Silicone spark plug cover

Compounded base, 45 hardness	250 lb
Amorphous silica	250
Red iron oxide, 50% in silicone fluid	5
Bis(*t*-butylperoxy)hexane, 50%	5
Coolant	Set at 80–100 °F, mixed at 20 rpm
0 min	Load rubber and color
1.5 min	Load silica slowly
3 min	Load peroxide dispersion
Dump	Dump at about 140–150 °F after 5–6 min
Mill	Cool; nylon or PTFE guides

Silicone polymers are not particularly elastomeric and they draw little power during mixing. As a result, only fine particle, easily dispersed fillers are practical, and trace additives must be used in the form of dispersions in silicone gum or fluid binders. Additives include colors and peroxide cross-linking agents.

The typical starting polymer has already been compounded by the supplier with reinforcing silica and coupling agent. This product is known in the trade as a **base**; the term is taken from the jargon of alkyd resins and phenolics (the original products of companies that now make silicones), in which a partly compounded composition is not, as the rubber chemist would say, a 'masterbatch', but a 'base'. The same suppliers refer to the totally uncompounded siloxane polymer as a **gum**, from the days when their uncompounded alkyds and phenolics were gummy indeed.

Siloxane polymers are usually so soft that the initial compounding with silica and coupling agent requires a dough kneader type of mixer. The rubber technologist most often starts with the compounded base, which is then extended and colored before the cure ingredients are added. Through the same unfortunate persistence of unrelated jargon, the cure initiators, usually organic peroxides, are known in the trade as **catalysts**, even though they are consumed during the cure.

As a consequence of the low internal strength of most silicone compositions, they are normally mixed with as little heat input as possible, i.e. with a relatively slow mix. Aside from traces of EPM or EPDM, contamination with other polymers and with typical compounding additives must be strictly avoided. Silicones are commonly mixed on dedicated lines, particularly for medical and other high end applications. Spark plug covers are one of the few end uses that can be accommodated, with a little care, by the general-purpose mixer. In the compound of Table 8.28, use of a base which would yield 40–45A hardness without extension provides 60–65A hardness in the final compound.

9

Mixing wire and cable compounds

Allen C. Bluestein

9.1 INTRODUCTION

In many respects the mixing of wire and cable rubber and plastics compounds is not very different from mixing compounds for other uses. In all cases the mixing process is often considered an art, and mixing wire insulation is considered to be a peculiar form of witchcraft. Few believe that scientific principles and fundamental physical phenomena are involved in every step of the process, whether or not they are understood by the many practitioners of the art [1–5].

Although there have always been attempts to understand, quantify and control the phenomena involved, there has been a substantial increase in these efforts in the recent past. The development of new materials and improved mixing equipment has been a more gradual process.

The typical wire and cable mixing process of the 1940s and 1950s involved the slow incremental addition of polymers, fillers, and process oils to a Banbury mixer running at 40 rpm, with tap water cooling on 'spray sides', a low pressure ram and low horsepower motors. The batch was mixed by time and temperature and was usually dumped when the operator-cum-artist decided it was well mixed. It was often dumped on a two-roll mill, peeled off in small rolls and conveyed by hand or wagon to a short-barreled extruder, where it was stuffed by hand and strained. The hot extrudate went through a breaker plate, forming what was descriptively called 'spaghetti'. The sticky strands of

The Mixing of Rubber
Edited by Richard F. Grossman
Published in 1997 by Chapman & Hall, London. ISBN 0 412 80490 5.

compound clumped together. Approximately 50 lb pieces were cut off and conveyed back to the mill, where they were milled, sheeted off by hand, cooled and stacked.

The process was not as inefficient as it may seem; this is because it was not uncommon to have base compound mix cycles of as much as 20 min, so there was time to get the batch off the mill, strain, get the batch off the mill again and even clean up a bit before the next batch was ready to dump. Because of the inability to control temperatures adequately, a common practice for adding the vulcanization system was to cut the base batch in half when this step was performed in an internal mixer.

9.1.1 Developments since the 1960s

Over the past several decades every wire company that processes extruded cross-linked products has become familiar with, and has probably taken advantage of, the development of chlorosulfonated polyethylene (CSM), nitrile rubber/PVC blends, cross-linkable polyethylene (XLPE), ethylene–propylene copolymers (EPM) and terpolymers (EPDM), and chlorinated polyethylene (CPE). Most companies are at least aware of the advantages and disadvantages of radiation cross-linking, thermoplastic elastomers and the new truly high voltage solid dielectrics. Everyone who uses continuous vulcanization processes uses tilted or catenary (rather than the original, horizontal) continuous vulcanization (CV) tubes. Many companies have learned to take advantage of vulcanization systems where cross-linking takes place in a molten salt or inert gas environment. These are comparatively sudden or dramatic developments compared with the evolutionary changes that have taken place during the same period. And perhaps curiously, it is the evolutionary changes that seem to have escaped attention. Few companies show much awareness of the possibilities and even fewer have successfully exploited them.

9.1.2 Some major changes

Table 9.1 lists some of the major changes in rubber and plastics mixing with typical internal Banbury type batch mixers [6]. Most of the changes lead to increased productivity from a batch process. But we should not ignore the development of continuous mixers and other advances that have contributed to a quieter, cleaner operation and longer-lasting equipment.

In spite of much effort and progress on the part of equipment suppliers to develop continuous mixers suitable for general-purpose mixing, it is still essentially a specialized field. Although the Farrel

Table 9.1 Major changes in rubber and plastics mixing

Old	Intermediate	New
Two-speed rotors		Variable speed
Low power (#11, 800 hp)		High power (1500 hp)
Low pressure ram (60 psi)		High pressure (to 100 psi)
Tap water cooling	Refrigeration	Tempered water
Spray side cooling	Cored sides	Drilled sides
Sliding door		Drop door
Spring tension seals		Hydraulic seals
Chromed surfaces		Alloy surfaces
Two-wing rotors	Four-wing rotors	New rotor designs
Mix until sounds okay	Power draw and temperature	Control over all variables

Corporation is certainly not the only manufacturer of continuous mixing equipment, they have probably done more work than any other manufacturer, and a review of their reports gives a thorough picture of the state of the art [7–14].

9.2 TEMPERED WATER

Improved cooling has been one of the most dramatic and underutilized advances in Banbury mixers over the years. Chamber sides, doors and rotors are cooled to maintain the highest level of shear during the mixing process. The old method of cooling the sides was to spray water between the walls of a double-walled chamber. Body temperature was controlled by manipulating valves by hand and measuring the temperature of the drainwater. Highly variable water temperature put excessive strains on the equipment, occasionally with disastrous results. The development of channels for cooling water strengthened the body, allowed the safer use of refrigerated water cooling, and ultimately led to the development of the drilled cooling channels of a modern Banbury.

The increase in heat transfer efficiency was so substantial that too much cooling actually decreased the efficiency when mixing some types of compounds. This observation led to the development of a closed-circuit cooling system, called a **tempered water system**, that provides control of the mixing surfaces at whatever temperature is considered optimum [15]. It has been suggested by some experts that this system is of no value without a mixer with drilled sides. For optimum control and benefits, drilled sides are a must. However, it has been suggested by others [16] that some benefits may be possible even with cored sides, but not with the old spray side design.

The following advantages are claimed for the system:

1. Shorter mix cycles by up to 50%
2. Reduced power consumption by 10–20%
3. Improved dispersion
4. Increased maximum batch size
5. Improved batch-to-batch consistency
6. Increased body life expectancy

The system itself is a carefully designed series of pumps and automatic valves, designed to develop controlled high volume flow of water across all the metal surfaces of the mixer. The mixing cycles are reduced by setting the temperature to the point at which the particular polymer or specific compound develops optimum shear conditions.

Figure 9.1 shows plots of the coefficient of friction versus metal temperature for some common elastomers [6]. High filler loadings tend to

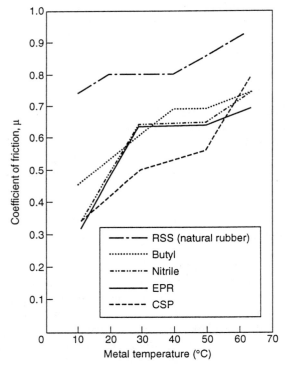

Figure 9.1 Common elastomers: coefficient of friction versus temperature.

Table 9.2 Optimum tempered water settings

Polymer	Starting temperature (°F)
Semicrystalline EPDM, MB	160
Natural rubber, MB	140
SBR, MB	140
Amorphous EPDM	100
Chlorosulfonated PE	80–100
NBR	80–100
Butyl (not heat interacted)	65–100
Neoprene	60–100
Curative additions[a]	80–120

[a]Will depend on activity.

increase the optimum temperature. Stocks that tend to build temperature quickly and are heat sensitive, such as black-filled polychloroprene (CR) compounds, require lower temperature settings than might be determined on the basis of shear alone. Optimum mixing temperature for each compound has to be determined experimentally, but Table 9.2 shows some practical starting points. The goal with heat-sensitive compounds is to use the lowest possible temperature that causes the polymer to grip the metal surface, enabling shear and turbulent flow of the polymer to take place, rather than slippage.

This goal is subject to the limitation that a special effort must be made to keep the metal surface temperature above the point at which atmospheric moisture condenses. But it is often difficult to avoid condensation when mixing NBR, butyl or polychloroprene in a humid environment since the best metal temperatures for these polymers seem to be in the range 50–75 °F, sometimes below the dew point. This effect is certainly a factor in 'summer mixing problems'. In the presence of condensation, filler dispersion will be inhibited and scorch safety and cure rate are sometimes affected. In addition, any moisture remaining in the compound could lead to porosity in the extruded wire covering.

With optimum mixer temperature the total power consumption is reduced because more time is spent mixing rather than moving the batch around. A greater fill factor is obtained because the mix is hugging the metal all the while it is in the mixer. Batch-to-batch consistency is improved because the temperature of the metal fluctuates in a very narrow range and each batch is exposed to essentially the same metal conditions at each step of the loading and mixing cycle. Improved

Table 9.3 EPDM UL 62, class 28 cord insulation: batch size for #11 Banbury

Semicrystalline EPDM, high Mooney	125 lb
Calcined clay	300
Paraffinic process oil	110
Zinc oxide	6
Antioxidant	1
Vinyl silane, 40% in paraffin	8
Dicumyl peroxide, 40%	16

Tempered water at 100–120 °F

Two-wing rotors: mixed at 25–28 rpm, conventional cycle

Four-wing rotors: mixed at 20–25 rpm, upside-down load

dispersion often manifests itself as an absence of unbroken-down polymer lumps. This effect can be dramatic with the mixing of highly filled, highly oil extended Underwriters Laboratories Class 28 EPDM insulations, such as illustrated in Table 9.3. It is difficult or impossible to mix compounds of this type efficiently with old-style internal mixers; even if one accepts the inefficiency, variable physical properties are often observed.

9.3 POWER-CONTROLLED MIXING

One of the most useful tools in studying the mixing process involves breaking up the curve of power draw versus time into zones characterized by a predominance of certain unit operations [26]. Note how this is similar to the analysis of power draw versus time used as a basis for contemplated changes to a given mix procedure (section 2.5). A typical curve is shown in Fig. 9.2. The first zone deals with loading time and a wetting stage, which involves the formation of a single mass of fillers and rubber, and the penetration of polymer into filler voids. It is during this part of the mix that the most time and energy is conserved by use of tempered water. The first zone is considered complete when the first power peak is observed.

The second zone is the area where most of the real dispersion work takes place. The filler agglomerates are gradually distributed through the polymer then broken down to their ultimate particle size. It, too, is terminated by reaching a power peak.

The third zone is where reduction of viscosity takes place. During this stage the rheological properties of polymers that are subject to

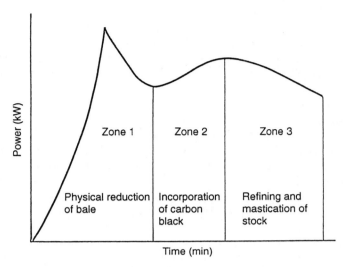

Figure 9.2 Zone analysis of upside-down power profiles.

molecular weight reduction by mechanical breakdown are modified significantly with increased mixing time. Even polymers that are not strongly subject to shear peptization undergo a disentangling process that reduces die swell and improves extrudability. There may also be some breakdown of very high molecular weight fragments.

The use of commercially available power integrators allows monitoring of the total power consumed at any point in the mixing cycle. This is certainly an improvement over using no power control, or working only with the total power consumed in the entire mixing cycle, but recent work indicates a need to view and control the entire power profile as well as the traditional time–temperature consideration [4]. One of the most comprehensive papers on this subject is by Gark and Sharman of Bayer's Polysar Division [27].

9.4 ENERGY CONSERVATION

Energy conservation, more scientific control of mixing conditions and machine criteria are all interrelated. Some of the processing conditions that conserve energy have already been discussed. Using the type of zone analysis described earlier, it is possible to develop power-controlled mixing techniques that not only conserve energy but also enable improved quality control. The use of power-controlled mixing has been reported to reduce energy consumption by up to 15% [18–21]. In one series of experiments, careful control of metal temperatures at a level that was considered to be optimum reduced total work by 40% [17]. Careful choice of batch size is often overlooked. Topcik has reported on a modified upside-down mixing procedure that reduced peak power by approximately 25% and total energy use by 10% [25]. The order of addition of the ingredients is also a topic that must be considered (examples may be found in Chapter 8).

Substantial savings may come from attention to machine parameters such as flow rate and control of cooling water, major elements of the tempered water system described earlier. Increasing the flow rate can sometimes reduce the dump temperature for a given energy input sufficiently to allow cure ingredients to be added in the first pass. The use of variable-speed motors allows more freedom to generate optimum shear rates at various stages in the cycle. Power override on the ram pressure control can prevent expensive power surges. Most later models of internal mixer are readily upgraded to incorporate these improvements (Chapters 1 and 5).

In the case of elastomers, subdividing the bales, even into relatively large (but less than bale-size) pieces, has been shown to reduce both peak power and total energy consumption by 10–15% [17]. Powdered rubber provides further reduction in energy consumption and may sometimes make the difference between a one-pass mix and a two-pass mix (section 2.4.1).

In order to reduce energy costs in the mixing process, it is essential to consider the choice of ingredients as well as processing conditions and machine parameters. Certainly the type and amount of filler must be considered. Some excellent work sponsored by J.M. Huber provides considerable data on this topic [22–24]. A number of additives, many of them metal salts of fatty acids, often blended with lubricants such as oleamide and erucamide, are specifically designed to reduce energy consumption. Recent work with *trans*-polyoctenamers has yielded similar results. Addition of relatively small amounts can sometimes improve the surface appearance of extrudates, leading to one-pass mixes with obvious energy savings and other economic benefits. Metal salts blended with lubricants can be particularly useful when polymer blends are used (e.g. neoprene and SBR).

REFERENCES

1. Klingender, R. C. (1977) Processing, in *Fourteenth Annual Rubber Group Lecture Series*, American Chemical Society, Akron OH.
2. Johnson, P. S. (1976) *The Incentive for Innovation in Rubber Processing – A Review*, CIC Rubber Group, Toronto, May 5, 1976.
3. Johnson, P. S. (1977) Processing of elastomers: a review of the problems. Paper presented at the Processing Symposium, ACS Rubber Division, Chicago IL, May 1977.

4. Johnson, P. S. (1979) Parameters in mixing, *Rubber World*, **April, May**.
5. Rapetski, W. A. (1974) The principles and practice of elastomer mixing. Paper presented at the Society of Rheology 4th Annual Meeting, October 1974.
6. Melotto, M. A. (1979) Machinery's contribution to better dispersion, *Elastomerics*, **111**.
7. MacLeod, D. (1975) Increasing Productivity of Older Size 11 and 11D Banburys, *Farrel Report C15–00231*.
8. Borzenski, F. J. (1975) Effect of Mixer Temperature Control on Batch Size, *Farrel Report C15–0061*.
9. Borzenski, F. J. (1975) Temperature Control Systems on a Banbury (SBR Mixing), *Farrel Report C15–0069*.
10. MacLeod, D. (1976) Recent Improvements in Rubber Mixing Equipment, *Farrel Report C15–0091*.
11. Borzenski, F. J. (1976) A Comparative Study of Performance of Batch and Continuous Mixers, *Farrel Report, Jan 1976*.
12. Drab, E. H. (1974) Continuous Mixer Designs and Applications, *Farrel Report*.
13. Melotto, M. (1978) Mixing, *Farrel Report C30–0081*.
14. Borzenski, F. J (1975) Influences of Higher Energy Cost on Rubber Mixing, *Farrel Report C30–00285*.
15. MacLeod, D. (1979) Tempered water provides improved mixing, *RPN Technical Notebook*, May 14, 1979.
16. Ellwood, H. (1977) Temperature control, *Eur. Rubber J.*, **Jan/Feb**.
17. Johnson, P. S. (1978) The challenge of energy conservation to the compounder. Paper presented at the Southern Rubber Group Meeting, Tampa FL, November 10, 1978.
18. O'Connor, G. E. and Putman, J. B. (1978) *Rubber Chem. Technol.*, **51**, 799.
19. Onnfer, R. J. (1966) *Rubber Age*, **98**, 51.
20. Van Buskirk, P. R., Turetsky, S. B. and Gunberg, P. F. (1975) *Rubber Chem. Technol.*, **48**, 577.
21. Dizon, E. S. and Papazian, L. A. (1977) *Rubber Chem. Technol.*, **50**, 765.
22. Barbour, A. L. (1976) Compounding to achieve lower costs, *Rubber Age*, **108**.
23. Barbour, A. L. (1978) Non-black filler/reinforcing materials for rubber, *Elastomerics*, **110**.
24. Pinter, P. E. and McGill, C. R. (1978) Comparing rubber fillers in an energy conscious economy, *Rubber World*, **Feb**.
25. Topcik, B. (1973) *Rubber Age*, **105**, 25.
26. Bluestein, A. C. (1979) Paper presented at the Regional Wire and Cable Technical Conference, Society of Plastics Engineers; *Elastomerics*, **119,** 10–15, March 1987.
27. Gark, K. and Sharman, S. (1989) Using a PC to monitor and analyze the processing of elastomers in an internal mixer. Paper presented at the ACS Rubber Division Meeting, Detroit MI, October 1989.

10

Mixing ethylene–propylene diene rubber

Charanjit S. Chodha and Emmanuel G. Kontos

10.1 INTRODUCTION

Not only are synthetic elastomers produced with a range of fundamental elastomeric properties so as to perform satisfactorily in service, but they are also designed to ensure that typical formulated compounds are mixed and fabricated into their final shapes with comparative ease. That is to say, synthetic elastomers now possess controlled processing properties.

The processing properties built into an elastomer may now be tailor-made to achieve easier and faster mixing and processing, controlled die swell, smooth surface appearance, etc.

In the case of EPDM, tailor-made processing properties can be achieved by controlling the average composition (ethylene/propylene ratio), the molecular weight and its distribution, the diene type and its level and, to some degree, the monomer sequence distribution.

Changes in the polymerization recipe and reaction conditions may require months or years of development work before they achieve the desired processing attributes while retaining or improving the fundamental elastomeric properties.

10.2 BACKGROUND OF EPDM DEVELOPMENT

EPDM elastomers were introduced some 30 years ago. At first, polymers were developed primarily to meet final product performance

The Mixing of Rubber
Edited by Richard F. Grossman
Published in 1997 by Chapman & Hall, London. ISBN 0 412 80490 5.

requirements. Little attention was paid to behavior during mixing and processing. It was left to the expertise of technologists at rubber product manufacturing companies to come up with working procedures for mixing EPDM compounds, either by empirical methods or by a more or less careful selection of polymer and compounding ingredients, together with systematic variation of processing conditions.

Over the years some notable improvements have been made to the polymer's macromolecular structure so as to improve its mixing and processing properties. At the same time, rubber equipment manufacturers have refined their equipment and control systems to enable the user to take advantage of state-of-the art elastomers.

No doubt many EPDM compounds are simple to mix, but the factors which govern their mixing are critical, including polymer selection, type of filler and level and Banbury fill factor. Such factors tend to be more significant for EPDM than for general-purpose elastomers, such as natural rubber, SBR and polychloroprene, when it comes to optimization of the mixing procedure.

10.3 COMPOSITION OF EPDM ELASTOMERS

Before giving detailed consideration to the mixing of EPDM elastomers and related parameters, it is essential to have a basic understanding of the polymer's composition, the effect of composition on its rheological properties and the ultimate effects on mixing.

In the presence of a catalyst – a transition metal compound and an aluminum alkyl halide – ethylene and propylene monomers are copolymerized to produce a high molecular weight rubber called EPM. EPM copolymers can be completely amorphous; they have a saturated backbone (i.e. no double bonds). A third monomer (a diene) can be included to provide unsaturation functionality along the polymer chain, creating the terpolymer (EPDM).

Typical molecular structures of EPM and EPDM are shown in Fig. 10.1. The degree of polymerization, denoted as m and n, may be quite large, typically about 2 000, and in polymers having elastomeric character m is about equal to n. The magnitude of the third monomer, y, is much smaller, often about 200.

The third monomer is introduced to enable sulfur cross-linking, but also enters into peroxide-initiated cure. Three dienes are generally used, all of them nonconjugated; they are ethylidene norbornene (ENB), dicyclopentadiene (DCPD) and 1,4-hexadiene (HD).

The ASTM nomenclature designating ethylene–propylene copolymer is EPM; ethylene–propylene terpolymer is EPDM. The M denotes a chemically saturated polymer chain of the polymethylene type, i.e. repeated $(-CH_2-)$ units in the backbone of the polymer.

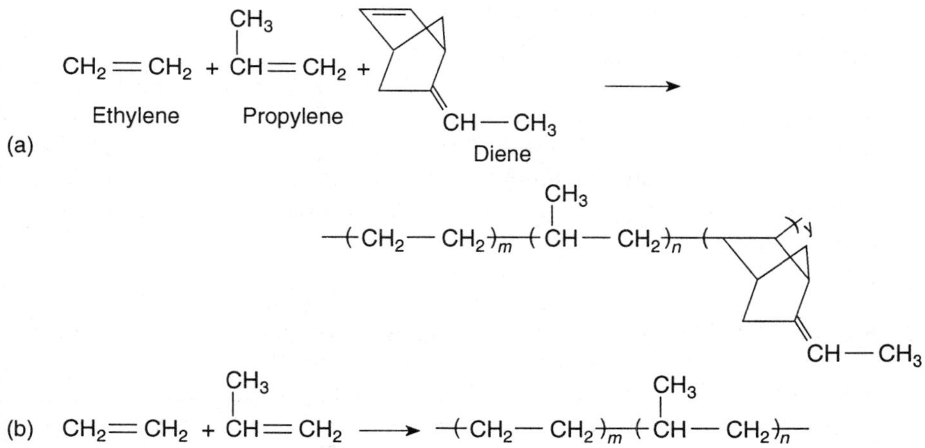

Figure 10.1 Typical structures: (a) EPDM terpolymer and (b) EPM copolymer.

All EPM and EPDM elastomers exhibit certain common properties such as greater resistance to ozone and resistance to higher temperatures than other general-purpose rubbers. And their ability to incorporate high levels of filler and oil generally leads to high cost-effectiveness. A particular application may require the correct choice of grade to maximize a combination of properties, maybe ease of fabrication, a specific vulcanization rate, high green strength, plus some set of physical properties in the vulcanizate.

10.4 VARIABLES IN EPM AND EPDM ELASTOMERS

Here are the key molecular parameters that can be varied in EPM and EPDM polymers:

- average molecular weight
- molecular weight distribution
- ethylene/propylene ratio
- type of termonomer(s) (EPDM only)
- level of termonomer(s) (EPDM only)

Now let us consider the above parameters individually and look at their effects on mixing, processing and, to some extent, on vulcanizate properties [1].

10.4.1 Average molecular weight

Increasing the average molecular weight increases the Mooney viscosity, it increases the green strength and it improves the collapse resistance, e.g. of the unsupported extrusions. Furthermore, it increases the possible filler and oil loading, thus reducing compound cost. Higher molecular weight, on the other hand, also tends to slow down filler dispersion, and can lead to problems with mill handling, extrusion and calendering.

Increasing the average molecular weight of the elastomer increases its tensile strength and tear resistance and reduces (improves) its compression set. As a first approximation, a high Mooney viscosity may be taken as a measure of relatively high molecular weight. But Mooney viscosity must refer to the base polymer. EPDM grades having substantial oil extension are based on high molecular weight polymers, but the product as purchased may have medium or even low Mooney viscosity.

10.4.2 Molecular weight distribution

Figure 10.2 shows different types of molecular weight distribution (MWD). Narrow MWD provides smooth extrusion, faster mixing and higher extrusion speeds, whereas broad MWD provides higher green strength and improved mill handling and calendering. A narrow MWD enables faster cure rates and correlates with improved low temperature compression properties.

10.4.3 Ethylene/propylene ratio

For most commercial rubbers the weight ratio of E/P ranges from 50/50 to 75/25. Increasing the ethylene content (which increases the crystallinity of the polymer) provides higher cold green strength, improves flow at process temperatures and enables high extension with fillers and oils. But there are limitations to increasing the ethylene content, such as greater difficulty in mixing and lower hot green strength.

The higher the ethylene content of the vulcanized elastomer, the higher its tensile strength, but to some extent it does sacrifice the compression set and the low temperature properties.

10.4.4 Type of diene

The type of termonomer has a strong influence on long-chain branching of the polymer, which in turn affects mixing and processing (section

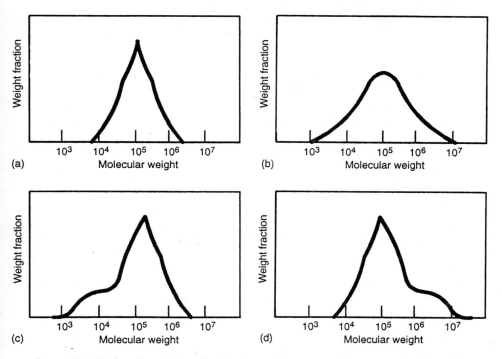

Figure 10.2 Molecular weight distributions: (a) narrow, (b) symmetrical, broad, (c) broad with low ends, (d) broad with high ends.

10.5). The degree of branching is highest with DCPD-EPDM and lowest with HD-EPDM.

Furthermore, the rate of sulfur-accelerator cure is slowest with DCPD-EPDM, higher with HD-EPDM and fastest with ENB-EPDM. With peroxide or electron beam-initiated cross-linking, the differences are quite small.

10.4.5 Diene level

Increasing the diene concentration has little effect on mixing and processing; any changes arise from the increased degree of branching. But a higher diene concentration does increase the rate of cure, it lowers the compression set and it provides higher modulus. Furthermore, it enables a greater range of accelerator selection, often useful in blends with other polymers.

Some limitations associated with high levels of diene are decreased resistance to heat aging, reduced scorch safety, decreased ultimate

Table 10.1 Characteristics of commercial EPDM grades

Designation	MW	MWD	E/P ratio	Diene	Diene level (%)
Royalene 521	low	medium	51/49	ENB	4.9
Royalene 502	med	narrow	62/38	ENB	4.0
Royalene 512	med	medium	68/32	ENB	3.8
Royalene 539	med	medium	74/26	ENB	4.5
Royalene 509	med	medium	70/30	ENB	8.0
Royalene 301T	low	broad	68/32	DCPD	3.0

elongation, higher polymer cost and a possible reduction in shelf life (bin storage).

Table 10.1 lists representative types of EPDM polymer, characterized as to molecular weight, MWD, E/P ratio, diene type and level. This listing is not meant to be exhaustive; typical polymer suppliers will be able to add further items to each category.

10.5 HOW PROCESSING RELATES TO STRUCTURE AND RHEOLOGY

Recent studies carried out by Beardsley and Tomlinson have increased our understanding of the mixing and processing of EPDM rubber in terms of its molecular structure and basic rheological properties [2]. The basic rheological properties are shear viscosity, viscous and elastic moduli and their ratio, tan δ (Chapter 14 gives a more general discussion of their effects).

For an EPDM having a given Mooney viscosity, the rate at which carbon black can be mixed into it increases as the molecular weight of the EPDM is lowered and as the degree of its branching is lowered. And the narrower the molecular weight distribution, the faster the mixing cycle. The reason that a high molecular weight, a broad MWD and a high degree of branching contribute to a slower mix is because of their higher elastic response (i.e. high elastic modulus and low tan δ values).

Higher degree of branching in a polymer is generally not desirable. Nonetheless, it is important to understand the effect of branching on processing because it plays a part in moderating cold flow and modifying other rheological characteristics. Grades that resist cold flow (e.g. bale deformation) are often characterized by the presence of numerous long branches. Branch entanglement at ambient temperatures significantly reduces flow. At the right level, and with branches of controlled size, this effect may be much more pronounced at room temperature, where it is useful, than at process temperatures, where it is a hindrance to flow.

Because of their partial crystallinity, polymers with high ethylene content must be thoroughly dispersed through heat and shear during mixing; this is to avoid unmixed lumps and the consequent poor dispersion. It often helps to warm the bales in a hot room. But the best solution for semicrystalline polymers is to use pellets or friable bales, which are easily broken into small crumbs as they hit the mixer rotors.

10.6 PRACTICAL GUIDELINES FOR MIXING EP ELASTOMERS

Given an understanding of how structure and rheology affect mixing, the next step is to examine how they relate to the equipment parameters. Efficient mixing of rubber compounds should not be the only goal; downstream processing and fabrication should also be considered.

How should one characterize a 'properly' mixed compound? Here are some of the significant observations and tests:

- Good or excellent polymer and filler dispersion (depending on the level required by the application).
- Consistent rheological and vulcanizate properties, both within a batch and, statistically, from batch to batch.
- Satisfactory processing on subsequent fabrication equipment.
- Absence of contamination to a degree required by the application.

Other criteria are mentioned in section 2.1.

10.6.1 Using internal mixers

There are several reasons why EPDM compounds are most commonly mixed in internal mixers:

- The mixing cycles are typically short.
- Internal mixers are convenient for adding high filler/oil loadings.
- The characteristics of many EPDM grades are best suited to use of internal mixers

Mixing in an internal mixer is accomplished in short mixing cycles, often 2–6 min. An upside-down mixing procedure is best suited for highly loaded compounds. With this technique (section 2.4.2) all of the compounding ingredients are charged into the internal mixer at the beginning of the mixing cycle, followed by the polymer. The curatives are typically added towards the end of the mixing cycle. A typical mixing cycle is given in Table 10.2.

Table 10.2 A typical internal mixing cycle (one-stage)

Banbury type 11D (236 L), water on, rotor speed 22 rpm	Time (min)	Temperature (°C)
● Load carbon blacks(s) + oil(s) + powders then the rubber(s)	0	50
● Raise the ram and scrape	1	75
● Add the curatives and brush down	4	100
● Dump	5	125

Table 10.3 A typical internal mixing cycle (two-stage)

Banbury type 11D (236 L) water on	Time (min)	Temperature (°C)
First-stage mix cycle (masterbatch): rotor speed 44 rpm		
● Load carbon black(s) + oil(s) + powders then the rubber(s)	0	50
● Raise the ram and scrape	1	90
● Dump	4	130
Second-stage mix cycle (final mix): rotor speed 22 rpm		
● Load ½ masterbatch + curatives then the other ½ masterbatch	1	50
● Dump	3	100

If the temperature in the internal mixer is high at the end of the cycle, the curatives may be added on the mill (Chapter 3). Alternatively, the first pass may be reloaded into the mixer and curatives added. A typical two-pass mixing cycle is given in Table 10.3. A two-stage mixing cycle is commonly 60% more expensive than a single pass and is therefore most often avoided. An important advantage of EPDM lies in the variety of compounds that may be mixed in a single pass (section 2.3 discusses the elements that affect the choice of single or multiple passes).

Efficient mixing depends upon several factors. Here are some suggestions to help achieve an effective short mixing cycle, with good polymer and filler dispersion, and with high consistency.

Polymer composition and form

Polymer type and physical form are significant. High ethylene EPDM grades can cause undispersed lumps during mixing because of the presence of crystallinity. To overcome this problem, use a friable bale (or pelletized grades) or prewarm the rubber.

Filler/oil levels and types

High structure and small particle size fillers are very difficult to disperse in ethylene–propylene rubber. An incorrect filler/oil level in the recipe can affect shear and temperature during mixing, hence reducing the quality of the mix.

Cure systems

When a fast sulfur or a peroxide cure system is used, it may be necessary to adopt a two-stage mixing cycle, or mix to a controlled low temperature in order to avoid onset of scorch during mixing.

Processing aids

No doubt some processing aids improve mixing and dispersion, but others can become incompatible during the mix, causing further processing problems or leading to surface bloom.

Mixing process

Mixing can be carried out using standard upside-down techniques (section 2.4.2) or with late addition of ingredients (section 2.4.1).

Mixing instructions

The mixing instructions should be simple and easy to execute. For greatest efficiency the maximum shear should be generated and maintained during as much of the mixing cycle as possible. The num ber of steps or additions to the mixing chamber is best kept to a minimum. Ingredients added in the wrong order will often lead to poor dispersion.

Fill factor

Improper batch size can lead to poor dispersion and consistency, whether it is oversize or undersize. Fill factor, generally 66–88% volume load, is selected by use of experience and observation to modify the basic considerations discussed in Chapter 1.

Mixing temperature

The mixing temperature depends upon power input, ram pressure, Mooney viscosity and filler particle size. For an active sulfur or peroxide cure system, the mixer temperature must be carefully controlled to avoid scorch or premature decomposition of the peroxide (section 9.2).

Machine parameters

Various rotor designs and friction ratios can seriously affect mixing efficiency (Chapter 5). Older internal mixers are usually equipped with single-speed or two-speed rotors (e.g. 22/44 rpm). Such mixers can be refitted with a variable-speed motor, enabling much closer control of shear and temperature during the mixing cycle.

Ram pressure

The purpose of applying ram pressure is to drive the ingredients into the mixing chamber. High ram pressure tends to increase the useful fill factor, as well as increasing filler–rubber contact, improving dispersion and increasing the ultimate mixing temperature. But higher ram pressure can accelerate wear on the mixer.

Coolant temperature

Almost every internal mixer that does not have a controlled temperature coolant system (popularly known as tempered water) may be refitted to include this significant improvement (section 9.2). Water temperature settings of 100–120 °F are often used with one-pass mixes and second-pass cure additions; settings of 150–180 °F are common with first-pass mixes, particularly with semicrystalline polymers.

Automation

Any automation of machine operation will tend to reduce the mixing cycle and improve the quality of the product by eliminating human errors. Whether or not these improvements can be realized will depend on the rigor with which automated components are maintained and tested (Chapter 7). Automation of supply systems (filler, oil, rubber, etc.) may be similarly effective in reducing the mixing time and eliminating any human errors; but vigilance is needed to avoid machine errors, which generally have a larger impact than human errors.

Machine condition

Excess wear will extend the mixing cycle, prejudice temperature control, interfere with dispersion and negatively affect the scorch time and cure rate. The precautions discussed in Chapter 12 are strongly urged.

Downstream processing equipment

Overhead mill blenders improve dispersion and quickly reduce stock temperature. The compound must have sufficient green strength to be processed through a blender. Multiple mills are sometimes substituted for an overhead blender. This alternative no doubt improves dispersion, but it may also reduce the stock temperature to a point where it becomes difficult to process. Instead of using a dump mill, the mixed stock is sometimes discharged into a ram extruder to produce plain or profile sheeting. Process temperatures in a ram extruder are normally higher than on a mill, so the stock must have a safe scorch time. This is less likely to be the case if the batch is dumped to a zero compression screw extruder with a short barrel. This type of sheet-off equipment, typically strongly water cooled and generally of high throughput, is most often called a **strainer extruder**, from its original purpose of screening critical compounds.

For best results, a 10–20% increase in volume over the calculated batch weight is recommended to achieve a good dispersion of filler and oil in extended polymer formulations. Try to minimize the number of additions during the mixing cycle (filler, oil, curatives, etc.), it is probably advisable to use a maximum of two or three increments, otherwise the additions will affect dispersion and prolong the mixing cycle.

The mixing cycle can best be controlled using a combination of measurements of time, temperature and power consumption, as discussed in section 2.5 (also section 9.3).

Several different mixing procedures can be used with EPDM formulations, but the simplest and most common procedure is the upside-down method. Other procedures common in industry are the late addition method, in which a part of the polymer, filler or oil is withheld for subsequent addition during the mixing cycle, and the incremental addition, or 'conventional' method in which the filler or the oil is added in portions during the mixing cycle. When using late or incremental addition, it is important that shear is not reduced substantially by multiple ingredient additions.

10.6.2 Using two-roll mills

Except for curative addition to Banbury-mixed first pass, little EPDM mixing is done on a two-roll mill. Where circumstances make mill mixing inescapable, the polymer selection is critical. The best EPDM grades for mill mixing are those with high propylene content (therefore amorphous), low Mooney viscosity and broad molecular weight distribution. In a typical procedure the rubber is loaded on a mill with a tight nip setting. As soon as the elastomer is banded on the mill roll, filler and oil additions are begun. There is no point in milling the uncompounded elastomer as, unlike natural rubber, it does not break down on milling. The compound should not be cut through while there are loose powders or oil because this will cause bagging. It is difficult to recommend specific mill roll temperatures because they depend upon the EPDM grade, filler/oil loadings, compound stiffness and the mill friction ratio. In mills with a friction ratio greater than $1:1$ (in practice, all except some mixer sheet-off mills), the stock will tend to band on the 'preferred' roll (regardless of roll speed or temperature). This may be either the front or the back roll. The best solution is to carry out mixing on the roll where the compound prefers to band. Attempting to avoid this usually wastes time in transferring the rubber from one roll to the other.

When all the fillers and oils are added, the compound may be cut down and blended thoroughly. There is seldom any need to cool the batch before the curatives are added. The curatives and the batch must be blended thoroughly so the stock is homogeneous. A typical mill mixing cycle is given in Table 10.4.

10.7 SUMMARY

Mixing EPDM rubber is simple and direct, but proper selection of the EPDM grade, filler/oil loadings and types, fill factor and mixing method are critical in achieving a homogeneously mixed compound.

Table 10.4 A typical mill mix cycle[a]

Roll length 60 in, water on, tight nip	Time (min)
● Load rubber(s) and band on the mill roll	0
● Add the carbon black(s) + powders + oil(s)	2
Band the stock and adjust the nip	
Return all the loose powders to the nip	
● Blend thoroughly	20
● Add the curatives and blend	27
● Take the stock off the mill	30
Cool the stock with water or air	

[a]Do not cut the stock when there is loose powder.

Subsequent processing and final properties must be built into the recipe to ensure that a product is manufactured efficiently with low scrap, and with the desired performance requirements. It is important that the structure and rheological properties of ethylene–propylene elastomers are well understood, so the factors which affect processing and vulcanizate properties may be properly addressed when formulating and mixing.

Current grades of EPDM are designed to overcome most of the problems associated with mixing, processing and fabrication. The days when a batch was scorched or prematurely vulcanized during mixing or processing should be long behind us. The degree of polymer and filler dispersion is still a concern because we have yet to develop a rapid means of taking measurements on-line. This is an area of active investigation.

REFERENCES

1. Chodha, C. S. (1989) A review paper in mixing ethylene–propylene diene rubber. Paper presented at the Chicago Rubber Group, October 2, 1989.
2. Beardsley, R. P. and Tomlinson, R. W. (1990) *Rubber Chem. Technol.*, **63**, 4.

11

Mixing of tire compounds

Martin J. Hannon

11.1 INTRODUCTION

Tire compounds are generally based on unsaturated diene polymers reinforced with carbon black, and vulcanized by an accelerated (usually a sulfenamide) sulfur system. The ingredients are normally mixed in a minimum of two steps. Polymers, carbon black, zinc oxide, stearic acid and any antiozonant or antioxidant system are put in the first step; curatives go in the second. The second step uses a lower temperature than the first step; this allows the stock (compound) to retain a window of processing safety. A typical first step starts with the polymers added at time zero, carbon black goes in at 30 s, and oil is added at 90 s; the dump temperature is 160 °C. A typical second step, or final step, has all the first pass and curatives put in at time zero, with a dump temperature of 110 °C.

The most important criterion for the compound is quality. A secondary concern is the amount of stock that can be processed in a short period of time. An average time for mixing a first step is 2–3 min, whereas the mixing time in the final step is typically 1–2 min. Maximizing throughput is important because a tire plant making 15 000 tires a day would have a daily requirement of 90 000 lb of tread stock, for example. Stock for the tread is the single largest compound used, but there can be as many as 10 other compounds needed to produce a tire. These include compounds for use with calendering steel belts, carcass fabric and innerliner sheets. Other stocks must be mixed for extruding treads, sidewalls and various reinforcing strips for the tire. The following

The Mixing of Rubber
Edited by Richard F. Grossman
Published in 1997 by Chapman & Hall, London. ISBN 0 412 80490 5.

sections cover four topics of particular interest to technologists mixing tire components; rework, phase mixing, viscosity reduction of natural rubber and the measurement of mixing efficiency.

11.2 REWORK

All factories generate rework, but very little is written about it. Most of the rework generated in a tire plant is produced during extrusion (e.g. treads too long or sidewalls too thick), especially at the beginning of an extrusion run. The physical properties of the stock have not been altered (greatly) by the extrusion process, but the extruded component cannot be used because its physical dimensions are not correct. There are several ways to rework the material back through the process. A mill convenient to the extruder may be designated for rework. Here all rejected extrudate is remilled and delivered to the feed mill supplying the extruder. This system is considered the least desirable because control of the amount of rework going to the feed mill (and therefore into the tire component) is difficult. A second possibility is to add a certain amount of rework to the mixer in the final step of mixing. This gives better control of the level of rework addition, but when adding multi-component rework (e.g. sidewalls and tread), it does not readily permit adjustment of chemicals or equalization of cost. A system that permits the use of multicomponent rework and equalization of cost is based on reformulation of the base compound, adjusting for the chemicals in the rework. Tables 11.1 through 11.4 illustrate this process.

Table 11.1 Base formula for compound A

NR	45.00
BR	28.00
1712	27.00
Carbon black	41.00
ZnO	2.20
Fatty acid	2.00
Wax	2.00
Antiozonant	3.50
Oil	8.00
Tackifier	2.00
Sulfenamide	1.00
Sulfur	1.50
Total	163.20

Table 11.2 Compound A in rework: rework strip 50% A, 50% B

	Parts	Percentage of formula	Parts of A in rework	Example
NR	45.00	27.57	3.60	(0.2757) (13.055) = 3.60[a]
BR	28.00	17.16	2.24	
1712	27.00	16.54	2.16	
Carbon black	41.00	25.12	3.28	
ZnO	2.20	1.35	0.18	
Fatty acid	2.00	1.23	0.16	
Wax	2.00	1.23	0.16	
Antiozonant	3.50	2.15	0.28	
Oil	8.00	4.90	0.64	
Tackifier	2.00	1.23	0.16	
Sulfenamide	1.00	0.61	0.08	
Sulfur	1.50	0.92	0.12	
	163.20			

[a]To add 16% rework, 26.11 parts of A/B strip are added to base formula. Half of the 26.11 parts come from A (13.055) and half from B (13.055).

Table 11.3 Compound B in rework: rework strip 50% A, 50% B

	Parts	Percentage of formula	Parts of B in rework	Example
NR	75.00	41.19	5.38	(0.4119) (13.055) = 5.38[a]
1712	25.00	13.73	1.79	
Blend of carbon blacks	61.00	33.49	4.37	
ZnO	3.00	1.65	0.22	
Fatty acid	2.00	1.10	0.14	
Antiozonant	1.90	1.04	0.14	
Oil	10.00	5.49	0.72	
Tackifier	1.00	0.55	0.07	
Sulfenamide	1.00	0.55	0.07	
Sulfur	2.20	1.21	0.16	
	182.10			

[a]To add 16% rework, 26.11 parts of A/B strip are added to base formula. Half of the 26.11 parts come from A (13.055) and half from B (13.055).

Table 11.4 Final formula with rework

Compound A		Compound A with rework	
NR	45.00	NR	36.02
BR	28.00	BR	25.76
1712	27.00	1712	23.05
Carbon black	41.00	Carbon black	33.35
ZnO	2.20	ZnO	1.80
Fatty acid	2.00	Fatty acid	1.70
Wax	2.00	Wax	1.84
Antiozonant	3.50	Antiozonant	3.08
Oil	8.00	Oil	6.64
Tackifier	2.00	Tackifier	1.77
Sulfenamide	1.00	Rework A/B	26.11
Sulfur	1.50	Sulfenamide	0.85
	163.20	Sulfur	1.22
			163.19

NR   45.00 − 3.60 (from A) − 5.38 (from B) = 36.02
N650 41.00 − 3.28 (from A) − (0.89 + 3.48) (from B) = 33.35

Depending on the overall amount of rework (16% in the example) added to the base formula, the amount of each ingredient in the rework is calculated, then those levels are subtracted from the corresponding ingredients of the base formula. Judgments sometimes have to be made, perhaps when the additives in question are not identical. For instance, compound B in Table 11.3 uses a blend of carbon blacks, compared to a single carbon black in the base formula. The technologist must decide what effects this combination of carbon blacks will have when it partially replaces the single black. This knowledge is generally available from previous compounding studies or can be gained from suitably designed experiments.

This system of planning for rework and accommodating its effects in the base recipe has been very successful in tire compounding. There are several reasons why: it allows more flexibility in extruding with multi-component rework; it keeps the variability of compound properties low; and it helps lower the cost of running the process. Planning for rework should be considered by all high volume compounders.

11.3 PHASE MIXING

Many compounding studies on phase mixing – the placement of filler or other chemicals in a specific polymer that is part of a multipolymer system – have been carried out in recent years, some of the most notable by Hess and coworkers [1]. One of their conclusions was that, in rubber blends, carbon black tends preferentially to reach equilibrium distribution in the polymer with the higher unsaturation, lower viscosity and higher polarity. For SBR, BR and NR blends, once carbon black becomes associated with a particular polymer, it tends to remain with that polymer. Table 11.5 shows data from the referenced study by Hess and coworkers. Of interest is the increase in tear resistance in an NR/BR system when the carbon black is forced into the NR phase. Many truck tire compounds use NR/BR polymer blends, and tear resistance is a concern, especially when the tires are used both on and off the road.

Table 11.6 shows a typical truck tire tread stock. Compound A used a normal mix; compound B was an attempt to put most of the carbon black into the natural rubber phase. A three-stage mix was used in which some of the carbon black was used to help oil addition in the second stage, but most of the black went into the first stage. The three-stage mix (Table 11.7) shows the expected reduction in viscosity, plus lower hardness, higher heat buildup under constant load and an improvement

Table 11.5 Properties of 50/50 polymer blends with N-220 type carbon black

Polymer system	Black addition	Resilience [a] (%)	T (°C)[b]	300% modulus (MPa)	Tensile strength (MPa)	Tear [c] energy (kJm^{-2})	Tread wear [d] resistance (%)
NR/SBR	To preblend	51.5	57.4	11.2	23.5	49.6	100
NR/SBR	75% in SBR	51.6	57.3	11.1	23.8	51.6	96
NR/SBR	75% in NR	54.2	53.3	12.1	22.7	33.4	95
NR/BR	To preblend	63.1	55.5	9.5	21.9	47.6	126
NR/BR	75% in BR	62.7	53.0	9.3	22.6	51.0	125
NR/BR	75% in NR	62.3	53.1	9.4	21.6	66.8	126
SBR/BR	To preblend	56.9	61.9	9.7	20.2	30.8	132
SBR/BR	75% in BR	58.6	61.6	10.3	19.7	26.4	139
SBR/BR	75% in SBR	57.3	62.0	10.1	19.2	33.6	133

[a]Goodyear–Healy rebound.
[b]Goodrich flexometer.
[c]Modified trouser tear test at 100 °C.
[d]Multisection recap treads, radial tires.

Table 11.6 Phase mixing NR/BR compound

Compound A		Compound B	
NR	50.00	NR	50.00
BR	50.00	Carbon black	40.00
Carbon black	55.00	ZnO	3.00
ZnO	3.00	Oil	7.00
Fatty acid	2.50		
Antiozonant	2.10	BR	50.00
Wax	1.50	Carbon black	15.00
Oil	14.00	Fatty acid	2.50
		Antiozonant	2.10
Sulfenamide	1.00	Wax	1.50
Sulfur	1.50	Oil	1.50
		Sulfenamide	1.00
		Sulfur	1.50

in tear resistance, especially hot tear. The higher heat buildup may be the price to be paid for the improved tear resistance. Tires were made using the compound B type of mix procedure and tested in the field. Results showed that this procedure indeed led to better tear resistance under field conditions.

Another conclusion reached by Hess and coworkers is that the hysteresis of an SBR/BR blend can be lowered by putting a major part of the carbon black into the BR phase. The data (Table 11.5) given previously on NR/BR blends suggested that this was the case. And compound B of Table 11.7 showed higher heat buildup when most of the carbon black was in the NR phase.

The lowering of tire hysteresis means lower rolling resistance, and this translates into more miles per gallon for a given automobile. Automotive manufacturers are considering every possible route to improved gas mileage; this is why evaluations were undertaken to measure the effect of phase mixing on SBR/BR compounds.

Table 11.8 shows formulations in which compound A is listed as a typical tread stock for passenger tires. Compound B is an attempt to locate more of the carbon black in the BR phase. BR leads to notoriously poor processing compounds when used alone. In practice this meant that some of the SBR content had to be mixed with the BR in the first stage. By placing most of the SBR in the second stage, the idea was to

Table 11.7 Physical properties of NR/BR compound

ML (100 °C) 1 + 4	*Compound A*	*Compound B*
	38	32
Mooney scorch at 135 °C		
Rise = 2 points	26	21
Rise = 10 points	28	23
Shore A hardness	63/64	60/60
300% modulus (MPa)		
20 min	9.17	8.58
40 min	9.63	8.58
Tensile strength (MPa)		
20 min	20.30	20.30
40 min	21.07	19.95
Elongation (%)		
20 min	495	560
40 min	525	550
Rheometer at 148 °C		
ML	9.6	7.7
MH	35.7	32.8
T_{s_2}	10.0	9.0
T_{50}	17.0	15.2
T_{90}	23.2	21.5
Heat buildup (°F) under constant load		
5 min	275	290
10 min	295	320
15 min	315	355
20 min	335	365
Goodrich flex (°F) under constant strain	55	55
Goodyear–Healy rebound (%)	64.1	63.1
Trouser tear		
M	3.6	4.8
D	3.7	5.2
Hot at 100 °C		
M	1.6	3.4
D	1.5	2.8

Table 11.8 Phase mixing SBR/BR compound

Compound A		Compound B	
Syn 1714	74.80	Syn 1714	37.40
BR	25.20	BR	25.20
Carbon black	63.10	Carbon black	63.10
ZnO	1.93	ZnO	1.93
Fatty acid	1.26	Fatty acid	1.26
Antiozonant	2.50	Oil	13.36
Oil	19.02		
		Syn 1714	37.40
		Oil	5.56
Sulfenamide	1.20	Antiozonant	2.50
Sulfur	1.60		
		Sulfenamide	1.20
		Sulfur	1.60

increase the carbon black level in the BR phase. Table 11.9 shows that Mooney viscosity, hardness and stress–strain data are similar for both compounds. The rebound was higher, and the heat buildup under constant load lower for compound B, indicating that the hysteresis of the compound should be somewhat lower. Tires were made and tested both with a wheel test and a fleet test in the field. Neither test showed any reduction in hysteresis. The stock mixed in the factory had similar properties to the lab mix, but, in practice, a major reduction in the hysteresis of tread stock is necessary to affect the performance of the tire and, in turn, of the automobile.

The balancing of one property of an elastomer blend versus another, to achieve an overall set of goals, invariably requires consideration and experimentation regarding which additives should be associated with which polymer phase. This is particularly the case with tire components, typically elastomer blends, but also applies to many other areas. It is extremely important to recognize that consistent properties in vulcanizates made from blends depend very strongly on using an identical mixing procedure with every batch. This may seem a simple enough proposition, and indeed has been axiomatic in the tire industry for many years.

11.4 NATURAL RUBBER: VISCOSITY REDUCTION

Tire companies continue to use large quantities of natural rubber. Radial passenger tires use a greater fraction of natural rubber than the older

Table 11.9 Physical properties of SBR/BR compound

	Compound A	*Compound B*
MS (100 °C) 1½	28	28
Mooney scorch at 135 °C		
Rise = 2 points	16	18
Rise = 10 points	17	19
Shore A hardness	58/60	58/60
300% modulus (MPa)		
15 min	7.07	7.70
30 min	8.89	9.87
Tensile strength (MPa)		
15 min	16.10	15.12
30 min	16.17	14.77
Elongation (%)		
15 min	565	495
30 min	485	410
Goodyear–Healy rebound (%)		
Room temperature	51.5	54.9
Hot, 100 °C	66.8	69.6
Goodrich flexometer, ΔT (°F)	57	62
Heat buildup (°F) under constant load		
5 min	270	215
10 min	280	225
15 min	285	230
20 min	285	230

bias ply tires; truck tires also use high fractions of natural rubber. A large portion of the natural rubber that is available has high Mooney viscosity (e.g. ML 1 + 4 at 100 °C $\geqslant$80) compared to common grades of various synthetic rubbers (which have a Mooney viscosity of 50–60 under the same conditions). To produce good mixing and processing compounds, the viscosity of natural rubber must be reduced. Several methods can be used. The natural rubber can be put into an internal mixer and physically broken down by the action of the rotors. In addition, chemical agents, such as peptizers or internally lubricating processing aids, may be added to the natural rubber in the mixer. Some natural rubber (e.g. CV-50) having controlled viscosity is also available.

Table 11.10 Mooney viscosity data ML (100 °C) 1 + 4 [a]

	SIR-5L				CV-50				CP				ILPA			
	BD	MB	RM	F	BD	MB	RM	F	BD	MB	RM	F	BD	MB	RM	F
NR at 0																
Rest at 90 s	–	111	–	94	–	85	–	73	–	86	–	65	–	97	–	83
NR, break-down	59	86	–	73	42	72	–	62	30	61	–	48	44	77	–	64
MB, remill	–	96	83	75	–	79	68	63	–	71	60	50	–	83	71	68

[a]BD = breakdown, MB = masterbatch, RM = remill, F = final.

The supply of controlled viscosity (CV) natural rubber is limited and there is a cost penalty that comes with its use.

For years chemical peptizers have been used to reduce the viscosity of natural rubber. There has always been some concern over how such peptizers affect the final properties of the compound. A few years ago the Pirelli Armstrong laboratory compared the mechanical breakdown of natural rubber with chemical viscosity reduction (peptizers and internal lubricating processing aids) and with controlled viscosity rubber [2]. The test compound was based on 100 parts natural rubber and 55 parts of N339 carbon black. This corresponded to a typical truck tire tread formulation. The natural rubber had a Mooney viscosity of 92. Table 11.10 shows the Mooney viscosity data comparison for the various treatments. The data indicate that chemical peptization leads to the greatest reduction in viscosity, whereas mechanical breakdown gives the smallest reduction. But Mooney viscosity data are always considered questionable because the shear rates associated with the Mooney viscometer are much lower than those actually developed in typical factory processes.

The Monsanto processability tester (MPT) is claimed to give viscosity values at shear rates that more closely approximate various factory operations. Data in Table 11.11 show the internally lubricating processing aid (ILPA) giving the most effective viscosity reduction at realistic shear rates. This is in contrast to Mooney viscosity testing, where the chemical peptizer (CP) appeared to be the most effective.

One of the concerns when using chemical additives is the extent to which they may affect the physical properties of the compound. Tables 11.12 and 11.13 show stress–strain data for the test compounds cured at two time cycles: twice and six times the value of T_{90} at 148 °C. Data at

Table 11.11 Monsanto processability test results: viscosity, K (Pa s)

	SIR-5L	CV-50	CP	ILPA
Shear rate = 55.3 s⁻¹ (calendering/milling)				
NR at 0, rest at 90 s	5.017	4.931	5.536	4.325
NR, breakdown	4.541	–	4.152	4.195
MB, remill	4.887	4.844	4.498	4.239
Shear rate = 99.7 s⁻¹ (calendering/milling)				
NR at 0, rest at 90 s	3.125	2.821	3.125	2.626
NR, breakdown	2.691	2.626	2.300	2.604
MB, remill	2.951	2.691	2.604	2.604
Shear rate = 299.1 s⁻¹ (hot feed extruder)				
NR at 0, rest at 90 s	1.344	1.156	1.329	1.113
NR, breakdown	1.170	1.098	0.982	1.098
MB, remill	1.214	1.105	1.105	1.055

Table 11.12 Stress–strain data for the test compounds

	SIR-5L		CV-50		CP		ILPA	
	2×	6×	2×	6×	2×	6×	2×	6×
300% Modulus (MPa)								
NR at 0	13.30	12.25	14.53	12.60	14.63	13.30	14.70	14.28
NR, break-down	13.48	12.60	14.04	12.81	13.65	12.32	14.63	14.42
MB, remill	13.48	14.43	13.65	12.04	14.35	12.78	14.07	13.65
Tensile strength (MPa)								
NR at 0	26.50	24.78	25.76	24.22	26.18	22.40	27.23	26.01
NR, break-down	25.55	24.61	26.39	23.66	25.13	23.38	25.55	25.80
MB, remill	25.76	24.36	25.62	23.52	25.48	23.10	26.78	26.53

Table 11.13 Differenced stress–strain data for the test compounds

	SIR-5L	CV-50	CP	ILPA
Δ 300% modulus, (2×) − (6×) (p.s.i.)				
NR at 0	1050	1925	1330	490
NR, breakdown	875	1225	1330	210
MB, remill	1050	1610	1575	420
Δ Tensile strength, (2×) − (6×) (p.s.i.)				
NR at 0	1715	1540	3780	1225
NR, breakdown	945	2730	1750	245
MB, remill	1400	2100	2380	245

Table 11.14 Various heat buildup tests

	SIR-5L	CV-50	CP	ILPA
Armstrong heat generation (°F)				
NR at 0	262	270	280	236
NR, breakdown	259	247	272	–
MB, remill	245	265	260	238
Firestone flexometer temperature (°F)				
NR at 0	249	255	260	239
NR, breakdown	251	264	257	246
MB, remill	245	251	242	248
Firestone flexometer blowout (min)				
NR at 0	18.1	16.1	12.8	22.6
NR, breakdown	17.6	15.5	14.3	20.4
MB, remill	16.7	16.0	16.4	22.2

twice 90% of full cure does not show much difference between the peptizing treatments, but Table 11.13 highlights the difference in properties between the double cure and the sextuple cure. The compound using internal lubricant gave the smallest difference (i.e. the most stable state of cure) whereas the other methods (chemical peptizer and CV-50) led to larger differences.

One of the reasons for natural rubber being used is to have a cool-running tire (low hysteresis). Data in Table 11.14 compare various tests that measure heat buildup. The compound with the internally lubricating processing aid always gives the lowest heat buildup. Based on the data generated in this study, use of the internal lubricant gave effective viscosity reduction with the lowest effect on the properties of the final compound.

11.5 MEASUREMENT OF MIXING EFFICIENCY

A tire factory generates large quantities of rubber compound, therefore mixing efficiency is a very important factor. Two very simple calculations for measuring the efficiency with which a compound is mixed will be discussed here. The first is called the stage mix formula (Table 11.15). It is simply the sum of the material parts in the first stage divided by the total recipe parts, plus the sum of the material parts in the second stage divided by total recipe parts, etc. The most efficient mix procedure would be a one-stage mix. Therefore, the closer one gets to a single-stage mix, the more efficient the process. This idea is discussed in section 2.3 for a compound totally unrelated to tire manufacture.

Another alternative is to measure the rate at which the mixer outputs compound, perhaps in pounds per minute. This formulation is shown in Table 11.15. Pounds per minute is equal to the weight of the final batch divided by the mix time (in minutes) of the final batch, plus a term, A. For a two-pass mix, A is the weight of the first pass multiplied by the time of the first pass then divided by the weight of the first pass.

In conclusion, the concept of formulation with rework in mind improves consistency as well as cost savings. The phase mixing concept

Table 11.15 Efficiency calculation

Stage mix

$$\text{Stage mix} = \frac{\text{material parts in first stage}}{\text{total recipe parts}} + \frac{\text{material parts in final stage}}{\text{total recipe parts}}$$

Mixer pounds per minute

$$\text{MP min}^{-1} = \frac{\text{weight of finish batch}}{\text{mix time of finish batch} + A}$$

$$A = \frac{(\text{weight of first stage in final}) \times (\text{first stage mix time})}{\text{weight of first stage}}$$

applies not only to fillers but also to plasticizers and vulcanization systems. Much potential exists for improvement of properties through better understanding of the distribution of additives in the compound. With natural rubber breakdowns, the data in the case of tire components clearly shows the advantage of processing aids. The two formulas for estimating efficiency may be quick and simple, yet they are effective in assessing the performance of mixing equipment and procedures. There is no reason why procedures developed to assist the survival of the tire industry cannot be adapted to the mixing of rubber compounds in general.

REFERENCES

1. Hess, W. M. and Wiedenhaefer, J. (1982) *Rubber World*, **186**, 15.
2. Herzlich, H. J., Hannon, M. J. and Daunais, J. W. (1983) Methods of reducing the viscosity of natural rubber. Paper presented at the Southern Rubber Group.

12

Mixing fluoroelastomer (FKM) compounds

Richard Mastromatteo

12.1 INTRODUCTION

Fluoroelastomers (FKMs) form a class of unique polymers, now enjoying increased use because of their extreme resistance to high temperatures and a variety of aggressive fluids. Although these elastomers use unique activator and cure systems, compounding variables are limited, and in some respects, fluoroelastomers are easy to formulate. The properties imparted by some of the compounding ingredients are extremely sensitive to moisture absorption as well as to particle size variation; these must be carefully controlled. Mixing is reasonably straightforward using internal mixers equipped with modern control systems (Chapter 1). Cleanliness is a key factor; contamination can have devastating effects. Because of the high cost of these polymers, careful consideration of the buy or make decision is very important.

The key steps in determining whether to buy or mix FKMs are to compare internal and external costs, then to consider the advantages and disadvantages, and finally to make the decision. Consideration of the costs of mixing FKMs is detailed in section 12.8. Thought should be given to the advantages of developing a partnership with a specialist having facilities and human resources dedicated to high performance elastomers because of the potential risk in handling such high cost materials in-house. In order to avoid some of the traditional pitfalls, the following guidelines are suggested:

The Mixing of Rubber
Edited by Richard F. Grossman
Published in 1997 by Chapman & Hall, London. ISBN 0 412 80490 5.

1. The decision should not be made by one department.
2. It is naive to assume that expensive and unique elastomers can be mixed more cheaply internally.
3. The purchasing department should initiate the analysis.
4. The best time to consider the decision is usually the introduction of a new product or new processing equipment.
5. The analysis must include all costs.
6. Overheads involved with mixing must be included, and can be very misleading.
7. A multifunctional team should make the final decision.

12.2 SPECIAL CONSIDERATIONS

Fluoroelastomers (FKMs) are polymers of the polymethylene type, with fluorine substituent groups on the polymer chain. These polymers are produced from vinylidene fluoride, hexafluoropropylene and tetrafluoroethylene monomers. Other fluorinated (or otherwise halogenated) monomers are used in small amounts for specialty grades, such as for improved low temperature properties, or introduction of peroxide-responsive cure sites. The resulting polymers have fluorine contents of 66–70 wt%. As the fluorine content increases, fluid resistance improves sharply. This is particularly the case with highly polar organic solvents, such as methanol, where volume swell decreases from 70% for 66% fluorine-containing FKM, to 3% for 70% fluorine-containing FKM. Fluoroelastomers are unmatched for performance in high temperature, low volume swell applications. They are designated by ASTM D 2000 as type HK, where heat resistance is rated at 250 °C, and volume swell in ASTM #3 oil is less than 10%. Other oil-resistant polymers are not in this class, with grades of NBR rated at 100–121 °C, and polyacrylates at below 150 °C. FKM elastomers form a class in themselves; one that is continually subject to increased demands by performance requirements. Current uses include O-rings, valve stem seals, shaft seals, gaskets and fuel hoses in automotive applications; O-rings, oven door gaskets and flue duct expansion joints in industrial applications; as well as oil well seals and packings and high performance roll covers.

Fluoroelastomer compounds are unique in several ways as compared with other synthetic rubbers. The compounds are fairly simple; they usually require only activator, filler and cure systems. Plasticizers are fugitive at typical service temperatures, so they are rarely used. Process aids are used only sparingly. Antioxidants and antiozonants seldom have sufficient heat stability to be of use in FKMs. The cross-linking agent most commonly used is bisphenol AF, accelerated with a phosphonium salt. Sulfur is not used and is, in fact, a poison for most FKM cure systems (Fig. 12.1). Grades of FKM bearing reactive sites (from specific

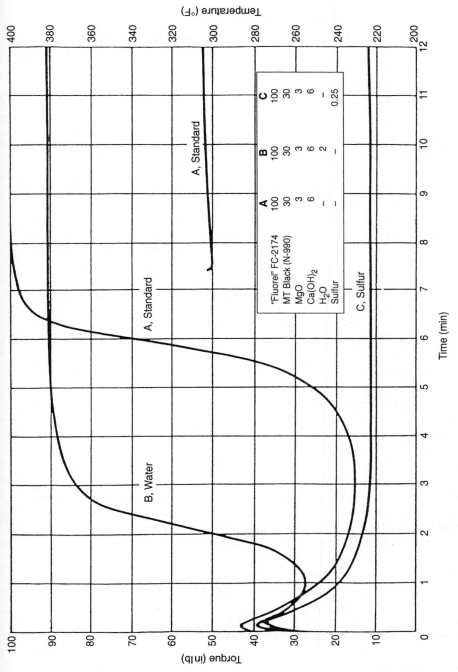

Figure 12.1 Effects of trace sulfur or water on the cure rate of a typical fluoroelastomer.

comonomers) are cured using organic peroxides. In some applications, particularly for bonding to other polymers, the oldest FKM cross-linkers, diamines (e.g. hexamethylene diamine carbamate and dicinnamylidene hexanediamine) remain in use.

12.3 RAW MATERIALS

The most important raw material is the elastomer (copolymer or terpolymer). With over 100 grades available, the compounder can usually find the optimum balance of heat, chemical and compression set resistance, and materials cost. Recent developments by polymer suppliers have created lower viscosity polymers having improved flow; cure systems that yield better compression set as well as fast cure speed combined with good scorch resistance; and grades which combine low mold-fouling characteristics with easy release. Parallel developments have improved the processability of 70% fluorine-containing FKM grades, and yielded clean-molding peroxide-curable polymers.

Additives are employed to optimize the properties of any elastomer. In this respect FKMs differ from most other synthetic rubbers. Compounds, perhaps based on NBR or EPDM, often contain a total of 10 to 15 ingredients (or more, see Chapter 2). But 5 to 8 ingredients suffice with FKMs, typically comprising polymer, curatives, acid acceptor/activators, filler(s) and a processing aid. As many grades of polymer contain incorporated cure systems and the use of processing aids is limited, compounders are mainly concerned with acid acceptors and fillers, the most common sources of problems.

The acid acceptors most widely used in fluoroelastomers are calcium hydroxide, calcium oxide, magnesium oxide, litharge (PbO) and dibasic lead phosphite. Calcium hydroxide is by far the most widely used activator, usually combined with high activity magnesium oxide. Important considerations when using these two ingredients include particle size (which contributes to activity level) and moisture content. Calcium hydroxide and magnesium oxide are hygroscopic, as is calcium oxide. Water has a strongly accelerating effect on the cure rate of FKMs, so it is imperative that raw materials remain dry. A convenient way is to use dispersions. Dispersions not only keep water absorption low, they also provide improved incorporation of the active ingredients. Achieving homogeneous dispersion of ingredients is potentially a problem, especially with low viscosity compounds. Particle size is of particular importance with litharge and dibasic lead phosphite, where small variations can produce significant rheological effects. Zinc oxide can also be used as an acid acceptor, and has found use in low viscosity applications.

12.4 TYPICAL FORMULATIONS

- Typical FKM test recipe

Curative containing polymer	100
Calcium hydroxide	6
Magnesium oxide	3
MT black, N-990	30
Carnauba wax	1

- Typical valve stem seal

Special-purpose grade	100
MT black, N-990	25
Calcium hydroxide	6
Magnesium oxide	3
Carnauba wax	1

- Acid-resistant compound

FKM terpolymer (68% F)	100
Litharge	15
MT black, N-990	30
Carnauba wax	1

- Water-resistant compound

Special-purpose copolymer	100
Dyphos	18
MT black, N-990	30
Carnauba wax	1

Fillers that are associated with tensile drift include talc, barium sulfate and ground silica. Inert fillers, such as iron oxide, do not display this phenomenon.

12.5 INTERNAL MIXING

FKM can be readily mixed with the same types of equipment used with other elastomers. Most commonly used is an internal mixer, such as a Banbury. Here is a list of process variables that must be controlled in order to mix FKM consistently:

- batch size (fill factor)
- loading procedure
- rotor speed (shear rate)
- rotor design
- ram pressure and design
- cooling temperature and flow rate
- sweep temperature

- drop temperature (indicated and actual)
- location of temperature indicator
- mix time

Batch size is determined by a formula (section 1.6.3) which contains a fill factor. In practice, the fill factor is a function of the ram pressure and the compound's specific gravity. The best indication of correct batch size is the behavior of the ram position indicator. If the ram occasionally bottoms out at the very end of the mix cycle, the batch size is correct.

The sequence and procedure for adding ingredients can affect the mix cycle (as well as the finished product). Ingredients that cause the batch to soften should be added as late as possible in the loading procedure. (On the other hand, if the batch on discharge is stiff, softening agents should be added earlier.) Ingredients that resist incorporation should be added first. With many formulations it becomes difficult to incorporate ingredients late in the cycle.

Varying the rotor speed allows more or less power to be delivered to the batch. The more power used, the faster the temperature will rise. The ability to use different rotor speeds permits control over the amount of power input. In order to maintain product consistency from batch to batch, the same power input must be used.

The standard two-wing Banbury rotor design has been very successful with FKM. The more recent four-wing rotor design will put more work into the batch in a shorter time, potentially improving productivity, but requiring close temperature control.

Several Banbury mixing techniques have been successful, depending on the type of compound and the application;

- one-pass mix
- two-pass mix
- upside-down mix (polymer added last)
- sandwich mix (powders between polymers)
- addition of ingredients to masterbatch

Chapters 2 and 8 give procedural details, especially for upside-down and sandwich mixing.

From the mixer, the batch would typically drop onto a two-roll mill. An important objective of sheet-off milling is to cool the batch quickly from mixing temperatures. A secondary purpose for this operation is to complete the mixing process. This is accomplished by rolling the compound into **pigs**, passed longitudinally through the mill nip, or by using a set of overhead blending rolls (Chapter 3). The compound is then cooled to room temperature and stored in a dry state.

12.6 MILL MIXING

Fluoroelastomers can also be mixed on a two-roll mill. The following is a typical procedure:

1. Premix all powders.
2. Set water to fully open for both mill rolls.
3. Set mill nip at ¼.
4. Band polymer on mill roll.
5. Add premixed powders across mill nip.
6. Sweep mill pan and add to batch.
7. Mix by overhead blender or by rolling pigs.
8. Remove material and cool.

FKMs must be kept contamination-free during the entire process. Depending on the compound and the application, preparation for a given mixing campaign ranges from use of a cleanout batch to partial disassembly and manual cleaning of the Banbury mixer. Because of material costs and the extent of cleaning needed to guarantee a contamination-free product, it is preferable to use a mixing line dedicated to FKMs.

12.7 SUMMARY

1. Always add soft materials to tough materials, never the reverse.
2. Compounds soften with increased temperature.
3. Add hard-to-disperse fillers to tough materials. Hard-to-disperse fillers include fumed silica, fine particle furnace black, calcium hydroxide and light calcined magnesium oxide.
4. Try to match viscosity when mixing:
   - Soft polymers can be made tougher with fillers.
   - Tough polymers can be made softer with temperature and plasticizer.
5. Be aware of the scorch safety of the compound. ODR T_{s_2} values below 2.0 min at 350 °F indicate sensitivity.

Unless temperature indication derives from an infrared thermometer (section 1.6.4), an instrument reading at the point of discharge will need to be adjusted, e.g. 220 °F usually means the batch is at 240–250 °F. A batch with a T_{s_2} of 2.0 min (ODR at 350 °F) probably has a 30 min life at 250 °F (Mooney scorch). This may not be sufficient to provide normal shelf aging followed by extrusion, calendering or molding. The first pass through the sheet-off mill takes a lot of heat out of the batch. This pass should be done relatively quickly.

Scorch times at 250 °F of greater than 10–15 min are generally adequate for extrusion of FKMs; molding, particularly of large parts, is better accommodated with 20–30 min scorch times. Values of less than 10–15 min indicate the probability of poor shelf life.

12.8 ACCOUNTING METHODS

The cost of FKM polymer is two to three orders of magnitude greater than for general-purpose elastomers (Table 12.1). With a specific gravity of 1.81–1.92, this cost differential is even greater on a pound volume basis. The cost factor is therefore very significant. Batch shrinkage during mixing could be costly. A batch lost to contamination can be extremely expensive. For example, if the value added during mixing is taken at $1.60 lb^{-1} and compound raw material cost is $13.50 lb^{-1}, it would be necessary to mix 8.4 batches of acceptable compound to balance the loss from one unrecoverable batch. If, to take a reasonable approximation, actual bottom-line profitability amounted to 10% of added value, it would require 84 batches to make up for the true loss. In other words, rejected batches simply cannot be accommodated by cost accounting procedures typical of mixing general-purpose elastomers.

Table 12.1 Analysis of compound costs

FKM base polymer	Price per pound ($)[a]	Cost for typical #3 Banbury batch ($)	Cost for typical #9 Banbury batch ($)
Peroxide curable			
66% fluorine	18.00	2 700.00	7 500.00
68% fluorine	18.00	2 700.00	7 500.00
70% fluorine	24.50	3 700.00	10 000.00
Cure incorporated: bisphenol cure			
66% fluorine	19.50	2 700.00	7 500.00
68% fluorine	19.50	2 700.00	7 500.00
70% fluorine	22.50	3 250.00	9 000.00
Specialty polymers: low temperature service grades			
66% fluorine	50.00	8 000.00	
69% fluorine	55.00	8 700.00	

[a]These prices are not expected to decrease in the near term.

The risk factor is even higher with specialty grades costing as much as $55.00 lb^{-1}. For example, at an added value of even $2.50 lb^{-1} during mixing, 17.4 good batches would be required to make up for the material lost on one bad batch, and 85 to recover the bottom-line profit. The bottom-line profit would amount to $739 500 worth of raw material, which would have to be committed just to break even.

13

Continuous mixing

Gene J. Sorcinelli

13.1 THE FARREL CONTINUOUS MIXER

The benefits of mixing rubber compounds continuously, rather than via a batch process, include the potential for improved statistical process control, because of the development of a steady state, and improved possibilities for automation. The first attempts to develop a continuous mixer for rubber date back to the 1950s, following many years of reliance on the Banbury internal mixer. This led to the introduction in 1960 of the Farrel continuous mixer (FCM), shown schematically in Fig. 13.1. It consists of a counterrotating twin-screw extruder having overall length to diameter ratio (L/D) of about 5. The feed zone is equipped with a hopper for metering a preblend of ingredients. This is followed by a mixing zone where the sections of the screws mimic Banbury rotors. The most common rotor design, featuring three mixing sections, is shown in Fig. 13.2. The final component is an adjustable discharge gate.

13.2 OPERATING PRINCIPLES OF THE FCM

The continuous mixer carries out the unit operations of mixing – incorporation, distribution, dispersion and viscosity reduction (section 2.2) – in much the same fashion as an internal mixer. Dispersion, the reduction in size of particle agglomerates, is achieved when the compound passes through the high shear zone between the rotor tip and the chamber wall, as shown in Fig. 13.3. The greater the number of times compound passes through this zone, the higher the level of dispersion.

The Mixing of Rubber
Edited by Richard F. Grossman
Published in 1997 by Chapman & Hall, London. ISBN 0 412 80490 5.

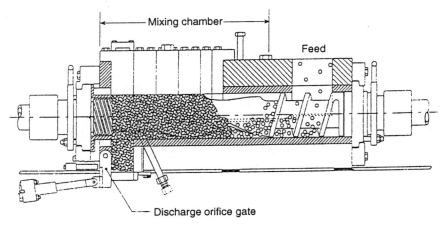

Figure 13.1 The Farrel continuous mixer: cross section. (Courtesy the Farrel Corporation)

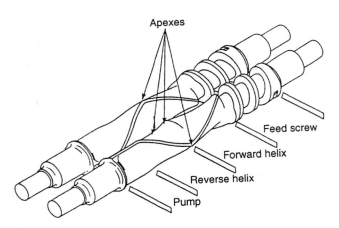

Figure 13.2 Rotor design for the continuous mixer. (Courtesy the Farrel Corporation)

Distributive mixing takes place in the area between the rotors, in the rolling bank of compound ahead of the nip formed by the rotor tip and the chamber, and from the back pumping caused by the reverse-helix rotor zone (Fig. 13.2). The extent of distribution is also governed by the number of times the compound passes through the high dispersion zone, as this fixes the number of times the same section of compound goes through the distributive lower shear zones.

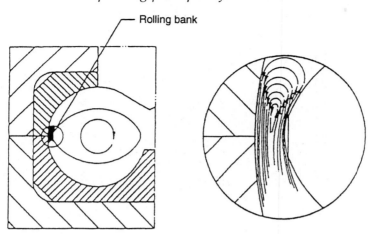

Figure 13.3 A high shear zone in the continuous mixer. (Courtesy the Farrel Corporation)

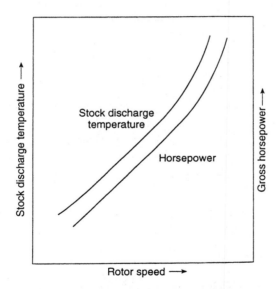

Figure 13.4 Effect of rotor speed on stock discharge temperature and gross power demand. (Courtesy the Farrel Corporation)

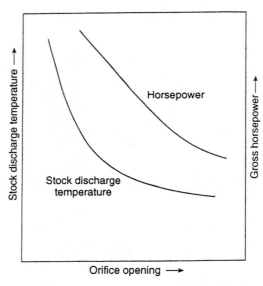

Figure 13.5 Effect of orifice opening on stock discharge temperature and gross power demand. (Courtesy the Farrel Corporation)

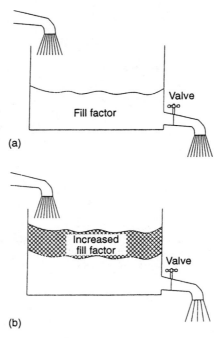

Figure 13.6 The water tank analogy: flow in and out for (a) initial fill factor and (b) increased fill factor. (Courtesy the Farrel Corporation)

Four variables control the operation of the FCM: production rate, rotor speed, discharge orifice setting and mixing chamber temperature. Since the continuous mixer operates at a steady state, the production rate, or output, is equal to the feed rate, or input. Increasing the feed rate cuts residence time, decreasing the extent both of distributive and high shear dispersion mixing.

The effect of rotor speed on compound discharge temperature and on power draw during mixing is shown in Fig. 13.4. At a given feed rate, and therefore residence time, increased rotor speed increases the extent both of distributive and dispersive mixing. But this is at the expense of increased energy input and, as with batch mixing, has the consequence of higher exit temperatures.

The effect of varying the exit orifice size on discharge temperature and power draw is shown in Fig. 13.5. As the exit opening is made smaller, at a given feed rate and rotor speed, power draw and stock temperature increase. This results from effectively increasing the mixer fill factor. A simple analogy with a water tank being continuously filled and drained is given in Fig. 13.6. A slight tightening of the exit valve produces a new point of equilibrium, with a higher level in the chamber equalizing output to input, but at higher pressure. Since a greater amount of compound is being sheared with each revolution the effect parallels that of longer residence time. Figure 13.7 summarizes the effects

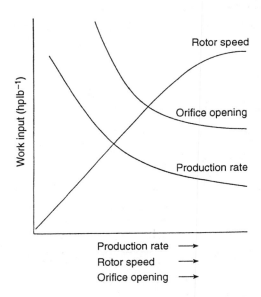

Figure 13.7 Work input per pound of material mixed: effects of production rate, rotor speed and orifice opening. (Courtesy the Farrel Corporation)

of production rate, rotor speed and size of exit orifice on work input per weight of compound.

The final variable, chamber temperature, affects the coefficient of friction of the compound, influencing incorporation and distribution as well as reduction of compound viscosity (section 2.2).

13.3 COMMERCIAL APPLICATIONS FOR THE FCM

FCMs range in size from the #1 LM (laboratory mixer), with an output as low as 5 lb h^{-1}, to the #18, rated at 80 000 lb h^{-1}. Output ratings of commercial mixers are given in Fig. 13.8. More than a thousand FCMs have been sold. Of these, only a small fraction, about 5%, are used directly to produce finished compound. Most of these are #4 LCMs used for one-pass mixing of mechanical goods. In a typical operation, a preblend of granulated rubber, fillers, pigments and curatives is fed to the #4 LCM at production rates of 300–600 lb h^{-1}. Excellent dispersion and consistency have been obtained under these conditions. Sometimes the output has been fed continuously to presses for what is, in essence, a continuous molding process from powder.

Larger LCMs have found a number of uses in the tire industry. These include final mix for tread and other compounds, often using a #9 LCM, for remilling and for warm feeding of compound to calenders. Use in these areas in place of internal batch mixers has been largely successful.

The major reason underlying the lack of widespread use of continuous mixing in the rubber industry has been the need for particulate

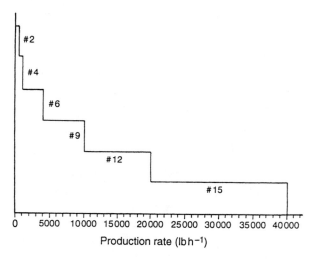

Figure 13.8 FCM size: effect on production rate. (Courtesy the Farrel Corporation)

rubber for metered feed to the equipment. The inclusion of pelletizing or granulating equipment ahead of the continuous mixer tends to increase overall equipment cost over that of a comparable internal batch mixer. Despite the potential for higher output, this has restricted capital investment in continuous mixing equipment. As a result, most FCMs are now sold to plastics processors for compounding of the range of thermoplastics having some elastomeric character. This equipment is ideal for mixing rubber/plastic blends, for example, as concentrates for improving the impact strength of thermoplastics, and for production of thermoplastic elastomers (TPEs).

13.4 THE FARREL MIXING VENTING EXTRUDER (MVX)

The unwillingness of industry to abandon bales in favor of particulate rubber led Farrel to develop the MVX mixer in 1975. Interestingly enough, this combination of internal mixer and extruder was designed for the plastics industry. But the plastics industry chose to adopt the LCM type of mixer, and the MVX is now used mainly by the tire industry. The basic design, a mixing chamber with rotors mounted above an extruder, is shown in Fig. 13.9. The size designation for the MVX consists of two diameters: the first is for the mixing chamber and the second is for the extruder screw; both of them are in millimeters. Current models include the following sizes: 134/120, 240/200, 360/250 and 565/300.

In Fig. 13.9 the feeding device (A) consists of a pneumatically operated ram which forces raw materials into the mixing chamber. This is quite similar to the ram in an internal batch mixer, but can be set to

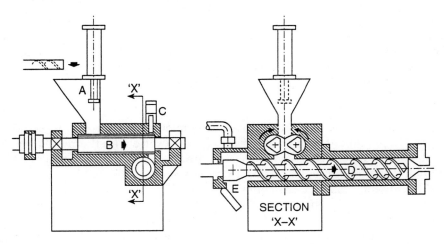

Figure 13.9 Basic MVX configuration. (Courtesy the Farrel Corporation)

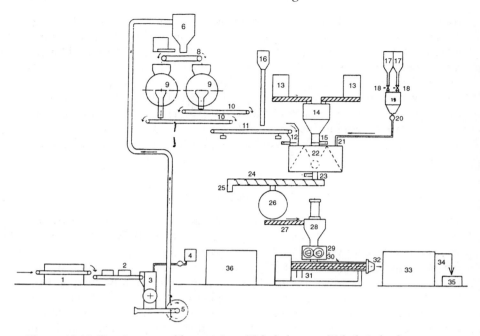

Figure 13.10 Continuous rubber mixing: (1) bale heater, (2) bale infeed conveyor, (3) bale granulator, (4) water/Tepol solution, (5) air conveyor fan, (6) cyclone, (7) antitack powder feeder, (8) rubber feed conveyor, (9) rubber storage drums, (10) rubber outlet conveyors to feed rubber weigh belts, (11) rubber weigh belts, (12) rubber inlet chute, (13) carbon black and powder day hoppers, (14) carbon black and powder weigh hopper, (15) inlet valve to premixer, (16) 'small powders' feed hopper, (17) oil storage tanks, (18) air-operated valves, (19) oil weigh scale, (20) oil injection pump, (21) oil feed pipe with injection nozzle, (22) premixer, (23) premixer discharge valve, (24) discharge conveyor, (25) divertor outlet, (26) storage bunker, (27) premix feedscrew, (28) MVX feed hopper, (29) MVX mixing chamber, (30) extrusion section, (31) reverse outlet, (32) extruder head, (33) festoon cooler, (34) wig-wag, (35) stack-on pallets, (36) computer control system. (Courtesy the Farrel Corporation)

reciprocate to drive a volumetric or gravimetric feed of raw material blend. Replacement with an auger is possible for use with thermoplastics instead of rubber. The MVX can accommodate much larger pieces of polymer than continuous mixing extruders such as the FCM.

The mixing chamber (B) contains two counterrotating, delta-shaped rotors. Tapers and helices are omitted since, as with the FCM, no forward pumping action is required. A variable speed drive operates the rotors at a 1.1:1 speed differential. Compound being mixed passes the length of the rotors and is driven through the exit port into the extruder.

Two provisions for venting trapped air, moisture or other volatiles are incorporated into the design. The mixing chamber is vented at atmospheric pressure (C) and the extruder equipped for vacuum venting (E).

The extruder section (D) can be of variable length, depending on the application. Shorter lengths are appropriate for simple strip used in subsequent processing; whereas the barrel and screw tend to be longer for direct extrusion of profile or other final shape. Although a straight delivery rubber extrusion screw is normally used, it is possible to add mixing sections if they are required.

Typical factory installation of the MVX is shown in Fig. 13.10. The general principles parallel the automation of any mixing facility. The most significant feature is the requirement for computer-controlled feedback to adjust the rates at which ingredients are delivered. This need is the only basic difference between an automated batch mixing installation and one that mixes continuously. Sometimes it is possible instead to automate a separate preblending operation for subsequent continuous feed to the mixer, as with the FCM, for example.

Table 13.1 gives the range of output for the various MVX sizes along with the typical power draw of the mixing and extruder sections. The mixing variables include rotor speed, feeding ram pressure, extruder screw speed, mixing chamber and extruder temperature. Output, or production rate, is governed by extruder screw speed. Input pressure and rotor speed must be adjusted to accommodate desired output while supplying sufficient mixing work to provide acceptable distribution and dispersion. Figure 13.11 plots mixing work, expressed in power draw per compound weight per unit time, versus rotor speed, feed pressure and output rate.

Independent control of the mixing and extruding sections of the MVX by means of separate drives has proven advantageous in several tire industry applications. These include single-pass mixing of bead, apex, sidewall and tread compounds based on synthetic elastomers, and condensation of three passes to two with natural rubber compounds (Chapter 11 reviews tire component batch mixing). There is a further

Table 13.1 Machine sizes (Courtesy the Farrel Corporation)

MVX size	Typical power (kW)	Typical outputs (kg h⁻¹)ᵃ
134/120	75/55	200/600
240/200	150/75	500/2200
360/250	450/225	1000/3000
565/300	1000/300	2500/5500

ᵃSpecific outputs depend on the compound being mixed.

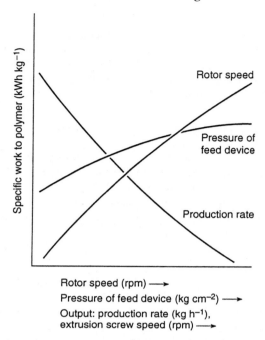

Figure 13.11 Work input per kilogram of polymer: effects of rotor speed, pressure of feed device and production rate. (Courtesy the Farrel Corporation)

advantage in the elimination of high energy peaks associated with the addition of rubber bales to an internal mixer.

Similar advantages have been obtained in the mixing of wire and cable components. As with mixing using the FCM, widespread commercial acceptance awaits the availability of a greater variety of elastomers in powder or pellet form.

14

Evaluating the performance of internal mixers

L. N. Valsamis, E. L. Canedo and G. S. Donoian

14.1 INTRODUCTION AND LITERATURE REVIEW

High intensity batch mixers trace their origin back to 1835, when the first roll mills were used by Edwin Chaffe for mixing rubber. The Banbury high intensity batch mixer followed in 1916. Many design changes to these basic rubber mixing devices have been implemented, particularly to improve the efficiency of dispersion of carbon black. But the underlying principle of operation remains unchanged. It is the repeated passage of compound through regions of high shear stress, followed by intimate mixing of the sheared material within the batch. Because of torque and heat transfer limitations, only a small portion of the compound is subject to high stress at any one time, but the overall rotor configuration ensures that the entire compound passes repeatedly through the high stress region. With a two-roll mill this process is accomplished by manually cutting and rolling parts of the batch for feed into the nip (Chapter 3).

In a high intensity batch mixer, the high stress region lies between the rotor tip and the chamber wall. This is the analog of the nip between the rolls of a rubber mill. A great deal of work has been devoted to the elucidation of detailed flow patterns in these regions. Simplified models of flow in the rotor tip region of internal mixers have been presented by Bolen and Colwell [1] and others [2–7]. Despite simplifying approximations, these studies provide some fundamental understanding of the shearing action in the high stress region.

The Mixing of Rubber
Edited by Richard F. Grossman
Published in 1997 by Chapman & Hall, London. ISBN 0 412 80490 5.

Another approach taken to identify overall flow patterns in batch mixers involves static flow visualization studies, such as described by Freakley and wan Idris [8], in which the mixer is stopped at an intermediate stage and the contents removed for examination. Similar studies by White and coworkers [9,10] described flow patterns in a laboratory mixer having transparent walls. Freakley and Patel [11] and Toki *et al.* [12] used pressure transducers to track the mixing process and to analyze flow near the rotor wings.

In recent years, advanced numerical techniques have been used to provide complete hydrodynamic analyses of internal batch mixers. In particular, the studies of Kim *et al.* [13] and of Cheng and Manas-Zloczower [14–16] developed global flow patterns for the entire mixer. The latter work showed good correlation between predicted and observed pressure rise in the region of the rotor tip.

The few studies that simulate carbon black dispersion in rubber in Banbury type mixers include those of Manas-Zloczower and coworkers [17,18]. Despite simplifying assumptions in the mathematical models, good agreement between theory and experiment was observed. A scale-up criterion for agglomerate dispersion was introduced by Manas-Zloczower and coworkers [19,20] based on the dimensionless product $X \cdot K$, where K is the average number of passes of compound over the rotor tip and X is the fraction of carbon black agglomerates broken during one pass. This parameter was compared to scale-up criteria of unit work input and total shear strain suggested by Van Buskirk *et al.* [21], Newell *et al.* [22] and Funt [6]. It was concluded that the $X \cdot K$ product applies over a broader range of mixer size and operating conditions. But the parameter X is based on a simplified representation of an agglomerate consisting of two touching spheres of equal size. And evaluation of X is cumbersome; it requires information regarding the nature of the agglomerate, its volume fraction, the diameter of the aggregate clusters within the agglomerate and the cohesive forces between them, and the viscosity of the polymer matrix under relevant shear rates and temperature. Many of them are not usually available for practical systems such as rubber–carbon black mixtures.

In subsequent papers, Tadmor [23,24] introduced the concept of the number of passage distribution (NPD) function as a means of characterizing dispersive mixing in batch and continuous mixers. The idea is based on the generally accepted theory that rupture of agglomerates in a polymeric mixture occurs through repeated passage over well-defined dispersion regions in the mixer. It does not require detailed knowledge of the actual mechanism. The complex and chaotic flow patterns within a high intensity batch mixer ensure that different sections of compound will experience a different number of passages through the high shear regions. The NPD function, the distribution of these passages, can be used to quantify dispersive mixing and aid in mixer design as well as batch scale-up.

The use of calculated parameters does not require details of the dispersive mixing mechanism, but only a description of the region of high shear. In most internal mixers this is the vicinity of the rotor tip and the tapered region upstream of the rotor tip. The tapered region induces elongational flow patterns that cause separation of sheared agglomerates [6]. This chapter considers the application of calculated parameters to commercial batch mixers along with the effects of rotor tip clearances. A new rotor design is also discussed and compared with standard two- and four-wing rotors.

14.2 DESIGNING THE ROTOR

The primary function of high intensity mixers is to incorporate more or less difficult-to-mix ingredients into rubber without more-than-desired effects on the polymer. This requires dispersion of complex agglomerates and provision of good distributive mixing with temperature uniformity. For example, it is well known that tire quality depends strongly on the level of carbon black dispersion. This must be achieved without overmixing and without otherwise degrading the molecular weight of the polymer beyond tolerable limits (section 11.4).

The rotors in batch mixers that did not intermesh, often called tangential rotors, traditionally ran at different speeds, commonly at a friction ratio of 13%. But it has recently been found that, with proper design and orientation, tangential rotors running at even speed can provide significant improvements [25,26]. The new Farrel design is called Synchronous Technology (ST). The design objectives were to increase productivity and uniformity while retaining versatility.

Figure 14.1 compares the top views of both kinds of four-wing rotors, standard and ST. The total helical and axial lengths of the four wings remains essentially unchanged, but the configuration of the ST rotor wings has been modified to provide a higher level of enforced order within the mixing chamber and to eliminate areas of poor flow. The cross section of the ST rotor has also been optimized to provide maximum extensional flow in the tip region while maintaining the shear stress levels of standard rotors.

The most noticeable change is that the two long wings (Fig. 14.1) now oppose each other, as compared to the standard rotor, where they drive compound flow unidirectionally. The effect is to improve compound flow with a double circulation, from the rotor center to each end. Together with the use of more open rotor area for greater axial flow, and redesign of the short rotor wings, the double circulation eliminates

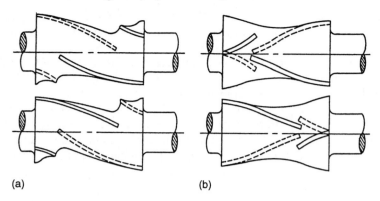

(a)   (b)

Figure 14.1 Rotor winds: (a) ST rotors have their long winds in opposition; (b) standard rotors have their long winds oriented for unidirectional flow. (Courtesy the Farrel Corporation)

quasi-stagnant areas at the rotor ends and proceeds most efficiently at even rotor speed. In conjunction with optimum alignment of the two rotors, an even rotor speed maximizes the squeezing flow patterns in the window of interaction of the two rotors and allows more efficient transfer of compound between the two chamber halves. Other design changes include increased free volume in a given mixer size, and improved temperature control of the two long wings of the rotor by circulating coolant to the region close to the rotor tip.

A similar design philosophy was applied to the development of the two-wing version of the ST rotor [26]. As in the four-wing design, the wing overlap was significantly increased to ensure that compound would traverse the entire mixing chamber and that a high level of enforced flow could be maintained. These conclusions were verified for both designs through flow visualization studies in a laboratory mixer [27].

14.3 ANALYSIS OF DISPERSIVE MIXING

As a first approximation, the complex flow pattern in the mixing chamber can be considered as the sum of two components: one in the axial direction, mainly responsible for distributive mixing; the other in the tangential direction, affecting mainly dispersive mixing. Characteristics of tangential flow discussed in the literature [1–7] include the presence of a rolling bank as well as a high shear zone in and around the rotor tip.

A simple Newtonian isothermal model has been developed to analyze the flow in the gap between the rotor wing tip and the chamber wall

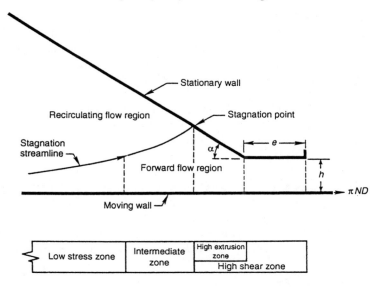

Figure 14.2 Schematic of the region between the rotor wing tip and the chamber wall. (Courtesy the Farrel Corporation)

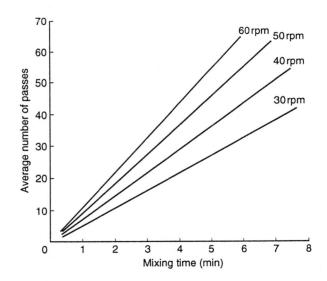

Figure 14.3 *K* value versus mixing time for different rotor speeds. (Courtesy the Farrel Corporation)

[28]. This region is shown schematically in Fig. 14.2, a linearized model of the wedge near the rotor tip clearance. In this diagram the rotor tip clearance is designated h, the land width e, and the leading wedge angle α. It is possible to calculate [28] the flow rate over the rotor tip per unit of axial length, q, without neglect of elongational stress, an important factor with elastomeric polymers [29]. This leads to calculation of the parameter K (section 14.1), the mean number of passes through the high shear zone in terms of time, t, the total axial wing length of one rotor, L, the net chamber volume, V, and the fill factor, ϕ:

$$K = 2\,q\,L\,t/V\,\phi$$

Figure 14.3 gives K values versus mixing time for various rotor speeds, using an F-270 Banbury with four-wing rotors at a 75% fill level.

The level of dispersive mixing is related to the maximum stress imposed on a section of compound, provided the maximum stress is applied for a sufficient time to break up agglomerates. Figure 14.4 shows

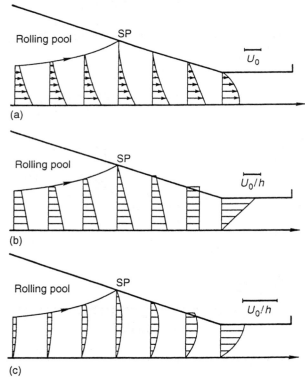

Figure 14.4 Dispersive mixing: (a) compound flow velocity, (b) shear rate and (c) rate of extension in the high shear zone. U_0: rotor tip velocity. (Courtesy the Farrel Corporation)

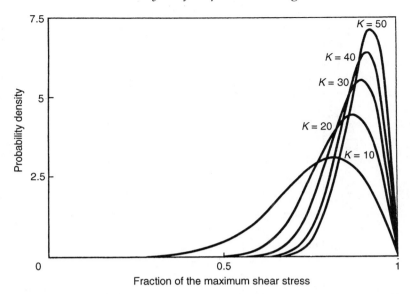

Figure 14.5 Distribution functions for maximum shear. (Courtesy the Farrel Corporation)

(a) the compound flow velocity, (b) the shear rate and (c) the rate of extension in the high shear zone. It is clear from this simple model that the shear stress imposed on the material passing over the rotor tip is highly inhomogeneous. During each pass, most of the compound travels near the moving rotor edge, where shear stress is at a minimum and residence time is short. Compound near the chamber wall experiences much higher stress and for a much longer duration. It is possible to calculate minimum and maximum shear rates for the tip clearance area and to develop a maximum shear distribution function [28]. This is illustrated in Fig. 14.5.

The maximum shear distribution function is used to predict the probability of the entire batch being subjected to the maximum shear rate during mixing. Figure 14.5 gives this probability density for five values of K, that are typical of rubber mixing, showing the effect of increasing the average number of passes through the high shear zone. As the value of the parameter K increases (more mixing), the fraction of compound receiving significantly more than the average shear rate in the high shear zone goes up exponentially. For example, at $K = 20$ more than 99.9% of the compound has been subjected at least once to shear rates higher than the average.

Another problem of interest is the heating of compound in the high shear zone, a potential cause of degradation. Here the non-Newtonian

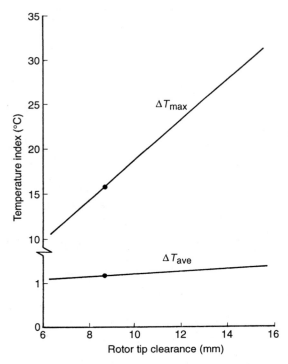

Figure 14.6 F-270 four-wing Banbury mixer: average and maximum local temperature increase in the high shear zone. (Courtesy the Farrel Corporation)

behavior of the compound, as well as the temperature dependence of physical properties, e.g. viscosity, cannot be neglected. Although solutions to these heat transfer problems are known, they are difficult to apply in actual cases [30,31]. Banbury mixers have relatively narrow tips and large clearances in which thermal transfer is far from equilibrium. And the region leading to the high shear zone is important, but is not considered in most treatments [32]. Nevertheless, a model has been developed for nonisothermal, non-Newtonian flow between nonparallel surfaces in relative motion [33]. Results to date suggest that, if an efficient cooling system is in use, average temperature increase in the high shear zone is modest, even for highly viscous elastomers under normal operating conditions. Substantially higher increases may occur locally.

A finding of the utmost importance is that local viscous heating increases sharply as the rotors wear and clearances increase. Figure 14.6 shows the average and maximum local temperature increase in the high shear zone of an F-270 four-wing Banbury, mixing a typical black-filled

EPDM compound at 60 rpm. Even though there is little effect on the average temperature rise, loss of specification clearances can easily produce localized overheating, with potential degradation of properties or development of scorch. This effect reinforces the need for regular inspection (section 7.1).

The results summarized in this section can be found in greater detail in Valsamis *et al.* [34]. Those with more extensive interest in the theoretical aspects of the mixing should consult the recent text by Manas-Zloczower and Tadmor [35].

14.4 RESULTS FROM SYNCHRONOUS ROTORS

The following study used a Farrel F-80 Banbury mixer equipped with two-wing ST rotors. The mixer was powered by a 600 hp DC drive with 4–140 rpm capability, the rotors running at even speed. A computer system was used, capable of operating the mixer in three control modes: batch temperature, total energy and constant mixing time. The batch temperature control program was used for one-pass mixing of the formu-

Table 14.1 Rubber one-step formulation (Courtesy the Farrel Corporation)

Component	Fraction (phr)	Fraction (%)
SMR-20 natural rubber	100	60.78
Sterling NS-1 (N-762 carbon black)	50	30.40
Sunpar 2280 oil	2.5	1.52
Zinc oxide	5.0	3.04
Stearic acid	2.0	1.22
Flectol H	2.0	1.22
Santocure	0.5	0.30
Sulfur	2.5	1.52
Total	164.5	100.00

Table 14.2 Mixing cycle sequence (Courtesy the Farrel Corporation)

Time = 0	Rubber, minors, ½ carbon black
45 s[a] or 180 °F	½ carbon black, oil
190 °F	Santocure, sulfur
220 °F	Brush
250 °F	Discharge

[a]Ram down time.

lation given in Table 14.1 according to the mixing cycle of Table 14.2. Batch temperature was measured by an infrared sensor in the doortop. Coolant temperature in the chamber, doortop and rotors was maintained at 90 °F by a Farrel three-zone tempered water system. Ram pressure was held constant at 50 psi with rotor speed set at 35 rpm. Total mix time, ram down time, batch temperature and power draw were recorded.

After drop, 10 temperature readings were taken at various points in the batch. After one pass through the mill, six samples were cut for Mooney viscosity and rheometer testing, two each from both ends and from the center of the batch. Mixer performance was checked at four different rotor orientations and for batch sizes of 45–65 kg. For each rotor orientation, at least four batch sizes were tested and, for each batch size, three batches of compound were mixed.

Cross-sectional views of three two-wing rotor orientations, corresponding to rotor alignments of 0, 90 and 180 ° are shown in Fig. 14.7. These alignments are based on the angle between the wing tips at the water (nondriven) end of the rotor. Because of the helix angle of the rotor wing, alignment varies along the length of the rotor. For this reason, Fig. 14.7 describes rotor cross sections of three axial locations, the two ends and the center. A top view of these rotor orientations is given in Fig. 14.8.

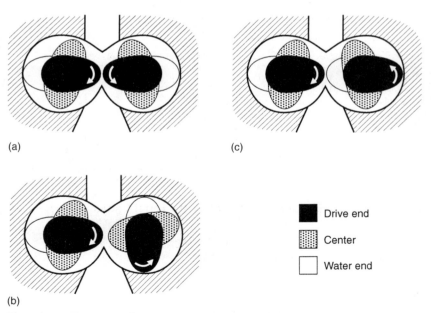

(a)

(c)

(b)

Drive end

Center

Water end

Figure 14.7 Two-wing ST rotors in cross section: (a) 0° orientation (b) 90° orientation and (c) 180° orientation. (Courtesy the Farrel Corporation)

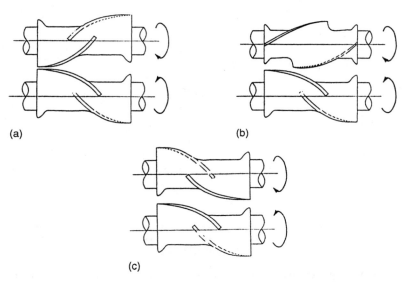

Figure 14.8 Top views of cross sections in Fig. 14.7: (a) 0°, (b) 90° and (c) 180°. (Courtesy the Farrel Corporation)

Figure 14.9 plots productivity in kg h⁻¹, assuming 100% running time, versus batch size at different ST rotor alignments. The alignment designated 0° gave the best overall productivity, followed by 180° then 90°. For each orientation there appears to be a different optimum batch size, with the 90° alignment having the greatest sensitivity. The effectiveness of the first two orientations appears to be controlled by the alignment of the wing tips as they interact. In the case of 0° alignment, the approaching wings pump in the same direction, effectively closing a portion of the window of compound flow and therefore causing compound to move back and forth along the axis. In the case of 180° alignment, even though the approaching wings are pumping in opposite directions, with minimal tip-to-tip alignment, they still effectively sweep the same window of compound flow, squeezing the compound between them. And with 90° orientation, the wings are completely out of phase; the compound flow between the rotors merely shifts from side to side as the alternating wings pass. This is the probable cause of the lower productivity and smaller optimum batch weight observed.

With a drop temperature of 121 °C, the 180° orientation consistently produced the smallest variation (<5°) in measured batch temperature, although all values lay between 120 and 127 °C. Figure 14.10 plots energy consumption versus batch weights for the different rotor alignments. The 0° alignment proved the most efficient and the 90° alignment the

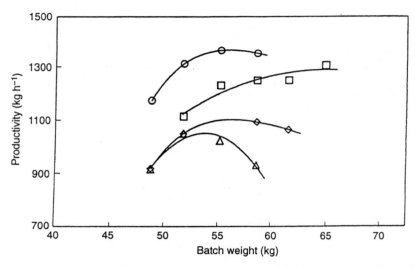

Figure 14.9 F-80 Banbury mixer with two-wing ST rotors: productivity versus batch size for (○) 0°, (△) 90°, (◇) 135°, (□) 180°. (Courtesy the Farrel Corporation)

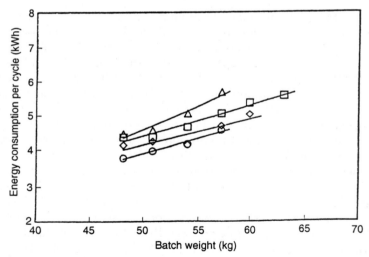

Figure 14.10 F-80 Banbury mixer with two-wing ST rotors: energy consumption versus batch size for (○) 0°, (△) 90°, (◇) 135°, (□) 180°. (Courtesy the Farrel Corporation)

Table 14.3 F-270 Banbury mixer productivity: ST versus standard rotors (Courtesy the Farrel Corporation)

Compound	Rotortype	Cycle time (min)	Mixing time (min)	Discharge temperature (°C)	Batch weight (kg)	Productivity (kg h⁻¹)	Change (%)
Neoprene	Standard	3.50	2.50	105	235	4028	
hose	ST	3.00	2.25	98	241	4820	+17
Fluoro-elastomer	Standard	3.75	3.15	110	380	6080	
seal	ST	3.50	2.90	105	380	6514	+7
Nitrile	Standard	3.25	2.90	110	226	4172	
gasket	ST	2.95	2.50	100	230	4678	+11
EPR	Standard	3.00	2.50	115	236	4720	
sheet	ST	2.75	2.25	110	242	5280	+11

Table 14.4 F-270 Banbury mixer properties: ST versus standard (Courtesy the Farrel Corporation)

Compound	Rotor type	Mooney viscosity		Maximum ODR torque		CB dispersion	
		Value	SD[a]	Value	SD[a]	Value	SD[a]
Neoprene	Standard	60	12.5	25.6	1.2	98.0	1.1
hose	ST	56	4.6	24.0	0.8	98.5	0.8
Fluoro-elastomer	Standard	75	6.6	28.6	1.8	99.0	0.9
seal	ST	72	3.8	26.5	0.9	99.0	0.6
Nitrile	Standard	55	8.9	24.5	1.1	97.5	1.3
gasket	ST	53	3.2	23.0	0.8	98.5	0.5
EPR	Standard	50	7.8	26.5	1.3	98.5	0.8
sheet	ST	48	3.4	25.2	0.7	99.0	0.7

[a]SD = standard deviation.

least efficient. This is surprising in view of the corresponding improvement in productivity with the 0 ° alignment. It may be that the change in rotor orientation affects the heat transfer characteristics of the mixer, through changes in compound flow patterns; the new heat transfer characteristic may lower the energy consumption. Rotor orientation, on the other hand, proved to have only minor effects on Mooney viscosity and rheometer torque.

Since the 180 ° rotor orientation yielded the most consistent batch temperature at drop, and followed the 0 ° orientation only slightly in productivity and power consumption, it was implemented in scale-up to an F-270 four-wing Banbury for comparison of ST with standard rotors. The test compounds were typical examples of neoprene hose, fluoroelastomer seal, NBR gasket and EPR sheet. Table 14.3 gives data on productivity using optimized batch sizes, and Table 14.4 gives data on compound properties. In every case the compounds could be mixed with higher productivity using ST rotors, with at least equivalent carbon black dispersion and comparable viscosity. The standard deviation in the measurements was also uniformly reduced.

This methodology for investigating the effects of using ST rotors can be generalized to the evaluation of a range of mixer design changes, or to compare different types of internal mixers [36].

REFERENCES

1. Bolen, W. R. and Colwell, R. E. (1958) *SPE J.*, **14**, 24.
2. Bergen, J. T. (1959) in *Processing of Thermoplastic Materials* (ed. E. C. Bernhardt), Reinhold, New York.
3. Mckelvey J. M. (1962) *Polymer Processing*, Wiley, New York
4. Guber, F. B. (1966) *Sov. Rubber Technol.*, **25**(9), 30.
5. Stupachecko, O. G., Pukhov, A. P. and Berbis, K. D. (1971) *Sov. Rubber Technol.*, **30**(7), 17.
6. Funt, J. M. (1977) *Mixing of Rubbers*, RAPRA Publications, Shrewsbury, UK
7. Tadmor, Z. and Gogos, C. G. (1979) *Principles of Polymer Processing*, Wiley, New York.
8. Freakley, P. K. and wan Idris, W. Y. (1979) *Rubber Chem. Technol.*, **52**, 134.
9. Min, K. and White L. J., (1985) *Rubber Chem. Technol.*, **58**, 1024.
10. Morikawa, A., Min, K. and White, J. L. (1989) *Int. Polym. Proc.*, **4**, 23.
11. Freakley, P. K. and Patel, S. R. (1985) *Rubber Chem. Technol.*, **58**, 751.
12. Toki, S., Takeshita, M., Morimoto, Y. and Okuyama, M. (1983) Paper presented at a meeting of the Rubber Division, American Chemical Society, Houston TX, October 24–28, 1983.
13. Kim, J. K., White, J. L., Min, K. and Szydlowski, W. (1989) *Int. Polym. Proc.*, **4**, 9.
14. Cheng, J. J. and Manas-Zloczower, I. (1989) *Polym. Engng Sci.*, **29**, 701.
15. Cheng, J. J. and Manas-Zloczower, I. (1989) *Polym. Engng Sci.*, **29**, 1059.
16. Cheng, J. J. and Manas-Zloczower, I. (1989) Paper presented at a meeting of the Polymer Processing Society, Amherst MA, August 16–17, 1989.

17. Manas-Zloczower, I., Nir, A. and Tadmor, Z. (1982) *Rubber Chem. Technol.,* **55**, 1250.
18. Manas-Zloczower, I. and Feke, D. L. (1988) *Int. Polymer Proc.,* **2**, 185.
19. Manas-Zloczower, I. and Tadmor, Z. (1984) *Rubber Chem. Technol.,* **57**, 48.
20. Manas-Zloczower, I., Nir, A. and Tadmor, Z. (1984) *Rubber Chem. Technol.,* **57**, 583.
21. Van Buskirk, P. R., Turetzsky, S. B. and Gunberg, P. F. (1975) *Rubber Chem. Technol.,* **48**, 577.
22. Newell, S. W., Porter, J. P. and Jakobs, H. L. (1975) *Rubber Chem. Technol.,* **48**, 1099.
23. Tadmor, Z. (1986) in *Integration of Fundamental Polymer Science and Technology* (eds L. A. Kleitjens and P. J. Lemstra), Elsevier Applied Science, Barking.
24. Tadmor, Z. (1988) *AIChE J.,* **34**, 1943.
25. Nortey, N. O. (1987) US Patent 4 714 350 (Farrel Corp.).
26. Nortey, N. O. (1988) US Patent 4 744 668 (Farrel Corp.).
27. Melotto, M. A. (1985) ACS Rubber Division, Cleveland OH, October 1985.
28. Canedo, E. L. and Valsamis, L. N. (1989) *SPE ANTEC,* **35**, 116.
29. Stevenson, J. F. (1972) *J. AIChE,* **18**, 540.
30. Rauwendaal, C. J. and Ingen Housz, J. F. (1988) *Intl. Polymer Proc.,* **3**, 3.
31. Lindt, J. T. (1989) *Polym. Engng Sci.,* **29**, 471.
32. Winter, H. H. (1977) *Adv. Heat Transfer,* **13**, 205.
33. Canedo, E. L. and Valsamis, L. N. (1990) *SPE ANTEC,* **36** 164.
34. Valsamis, L. N., Canedo, E. L. and Donoian, G. S. (1989) Methods in evaluating the performance of internal mixers. ACS Rubber Division, Detroit MI, October 1989.
35. Manas-Zloczower, I. and Tadmor, Z. (1994) *Mixing and Compounding of Polymers,* Hanser, New York.
36. Donoian, G. S., Canedo, E. L. and Valsamis, L. N. (1991) Optimizing mixing with ST rotors. ACS Rubber Division, Detroit MI, October 1991.

Appendix
Unit conversion tables and factors

A.1 CONVERSION BETWEEN FAHRENHEIT AND CELSIUS

Locate the temperature to be converted in the center column; read Celsius equivalents to the left, Fahrenheit equivalents to the right.

°C		°F	°C		°F
−56.7	−70	−94.0	104.4	220	428.0
−51.1	−60	−76.0	110.0	230	446.0
−45.6	−50	−58.0	115.6	240	464.0
−40.0	−40	−40.0	121.1	250	482.0
−34.4	−30	−22.0	126.7	260	500.0
−28.9	−20	−4.0	132.2	270	518.0
−23.3	−10	14.0	137.8	280	536.0
−17.8	0	32.0	143.3	290	554.0
−12.2	10	50.0	148.9	300	572.0
−6.7	20	68.0	154.4	310	590.0
−1.1	30	86.0	160.0	320	608.0
4.4	40	104.0	165.6	330	626.0
10.0	50	122.0	171.1	340	644.0
15.6	60	140.0	176.6	350	662.0
21.1	70	158.0	182.2	360	680.0
26.7	80	176.0	187.8	370	698.0
32.2	90	194.0	193.3	380	716.0
37.8	100	212.0	198.9	390	734.0
43.4	110	230.0	204.4	400	752.0
48.9	120	248.0	210.0	410	770.0
54.4	130	266.0	215.6	420	788.0
60.0	140	284.0	221.1	430	806.0
65.6	150	302.0	226.7	440	824.0
71.1	160	320.0	232.2	450	842.0
76.7	170	338.0	237.8	460	860.0
82.2	180	356.0	243.3	470	878.0
87.8	190	374.0	248.9	480	896.0
93.3	200	392.0	254.4	490	914.0
98.9	210	410.0	260.0	500	932.0

A.2 CONVERSION OF ENGLISH UNITS TO METRIC UNITS

Units of length	
inch	= 2.5400 cm
foot	= 0.3048 m
yard	= 0.9144 m
mile (US)	= 1.6093 km

Units of area	
in^2	= 6.4516 cm^2
ft^2	= 0.0929 m^2
yd^2	= 0.8361 m^2
square mile (US)	= 2.5900 km^2

Units of volume	
in^3	= 16.3871 cm^3
ft^3	= 28.3169 liters
yd^3	= 0.7646 m^3
pint	= 0.4732 liter
quart	= 0.9464 liter
gallon (US)	= 3.7854 liters

Units of mass	
ounce (avoirdupois)	= 28.3495 g
pound	= 0.4536 kg
US ton (2000 lb)	= 907.20 kg
2204.62 lb	= metric ton (1000 kg)

Common units for rubber	
Adhesion (lb/linear in)	= 0.1751 kN m
Stress (psi)	= 0.006 895 MPa
Torque (in-lb)	= 0.1130 N m

Industrial units	
horsepower (hp)	= 0.746 kW
BTU	= 1054.62 J
small calorie	= 4.185 J
atmosphere	= 1.0333 kg cm^{-2}

Index